GO TOXIC FREE

GO TOXIC FREE

Easy and Sustainable Ways to Reduce Chemical Pollution

Anna Turns

Michael O'Mara Books Limited

First published in Great Britain in 2022
by Michael O'Mara Books Limited
9 Lion Yard
Tremadoc Road
London SW4 7NQ

A CIP catalogue record for this book is available from the British Library.

Papers used by Michael O'Mara Books Limited are natural, recyclable products made from wood grown in sustainable forests. The manufacturing processes conform to the environmental regulations of the country of origin.

ISBN: 978-1-78929-343-2 in hardback print format
ISBN: 978-1-78929-344-9 in ebook format

1 2 3 4 5 6 7 8 9 10

www.mombooks.com

Designed and typeset by Ed Pickford

Printed and bound by CPI Group (UK) Ltd, Croydon, CR0 4YY

Dedicated to the eco-conscious citizens around the globe
who all play a part in the fight for pollution justice.

Contents

Introduction

Awareness of plastic pollution has skyrocketed in recent years. But what's inside those bottles, packaging and containers, and what else besides plastic is our stuff made of?

The 'useful' lifespan of a cleaning spray, shampoo or toy is just a fleeting snapshot of a much bigger story, and it's really important that we understand what happens along these hidden and complex supply chains. It's crucial that we consider the ingredients and raw materials that are used to make the contents of each bottle, carton or can that we buy, and what happens when they get thrown away after we've finished using them. Often, hundreds of different chemicals are involved in each production process, and many of them are not just harmful to us while in our homes, but poisonous to the planet before and after we use them.

From the chromium used in commercial leather tanning that leaches into the waterways to the mercury that poisons small-scale gold miners, every toxic chemical has a knock-on effect somewhere along the line, either as an environmental contaminant or a human health concern; sometimes both. So the domestic choices we make today about what we buy, how we use it and what we do with it afterwards all have far-reaching impacts.

While chemical pollution is a global problem, sometimes it can occur in places where you'd least expect it. Everything is connected. We are all part of this Earth's ecosystem; not above it and definitely not separate from it. Our actions affect our environment and each other, as well as future generations, here and on the other side of the world. The good news is that we can do so much to minimize our own chemical footprint.

This journey starts at home, it goes global and it keeps coming back to how we can reduce chemical pollution in our daily lives. In the first part of this book, I explain what a chemical is, what chemical-free isn't and what toxic really means. I shine a light on the science so that you can see through the greenwash and discover which claims are myths and which hold truth. By lifting the lid on toxics, my own assumptions have been challenged massively and there have been plenty of surprises along the way. Advertising culture skews our perceptions, so be prepared for your understanding of what's safe and what's not to be challenged. And the worst culprits aren't necessarily the ones we hear about all the time, so I've created a list of top toxics which highlights the things that we need to focus on most of all.

While unpicking the issues surrounding our use of potentially harmful chemicals, I've found that exciting solutions already exist and plenty more are on the horizon. So, these pages are full of progress, innovations and change-makers who are already making headway, from remediation at the source of pollution to the design of more sustainable chemicals that won't harm the environment.

By joining the dots between us, our families and pets, our homes and the surrounding environment, we can all tread more lightly. Our carbon footprints, plastic footprints and toxic-chemical footprints can be reduced. But zero impact is meaningless because every single thing we buy, eat, spray and use on our bodies has an environmental knock-on effect. Yes, we can streamline the number of toxic chemicals that we use and buy products that are less toxic. But it's possible to go one step further. By buying substances and materials that have been formulated in a positively regenerative way or changing our habits altogether, our choices can actually enhance the environment. Instead of sourcing fewer toxic pesticides for your garden, ditch them altogether and create a wildlife-friendly habitat where biodiversity thrives and the food you grow is free from chemical residues. Every action has an impact, but it can be a positive one.

Room by room, I investigate which things in our homes pose the most risk and which we could replace with healthier alternatives or different routines altogether. I discover the exact components of my house dust, I get my blood tested for the existence of toxic chemicals and I find out what happens to the air quality in my home when I burn the toast. My mission is to make the reality of chemical pollution and what we can do about it as tangible as the plastic litter we might see on the street, verge, riverside or beach.

From what gets washed down the plughole in the kitchen to what might get flushed down the loo, it's clear that everything has a consequence, however small. There is no washing or flushing 'away'. Think of Part Two as a guide around your home. Feel free to dip in and out, and revisit whenever you need to and find what works best for you. Perhaps focus on the cupboard beneath the kitchen sink or check out what's in the detergents you use on your family's clothes, bedding and towels. Create your own recipe for a healthier home. Try alternative solutions, tweak ideas, experiment with possibilities and adapt your lifestyle.

Decisions aren't guaranteed to be clear-cut. There isn't always a right or wrong option. This book will arm you with the relevant facts and the latest thinking from experts around the world to build a more balanced picture. Sometimes, you might end up asking numerous questions. I urge you to be curious, spark new conversations and push for more answers. Email manufacturers to ask for further information about claims on their labels or for lists of ingredients that haven't been published yet and support independent, ethical businesses aligned with your own values. One thing's for sure – by demanding greater transparency, we can be catalysts for real transformation.

PART ONE

The Big Picture

1

The Clean, Green, Toxic Machine

Eco-friendly. Chemical-free. Natural. That's all great stuff, right? Wrong. There's so much confusion about what's toxic and what's not. Let's start with a simple definition – what is a chemical? It's just a substance. Our world is made up of millions of chemicals, most of which have complex scientific names. Every ingredient on this planet is a chemical. The oxygen we breathe is a chemical. So is water. We need chemistry to survive. In 2010, the Royal Society of Chemistry offered £1 million to the first person in the world to create a chemical-free product. No one has yet claimed the bounty because it's an impossible task.

What's clear is that words really matter. The term 'chemical-free' means nothing, while eco-friendly could imply one of many things. Many unregulated terms are used on labels by companies to advertise their stuff and it shouldn't have to be our job to decipher them. So let's start by debunking some myths.

Naturally confusing

First off, toxic free means free from harmful or hazardous chemicals. Anything can be toxic at high enough concentrations, even water or oxygen. Generally, the dose makes the poison, as the sixteenth-century physician and alchemist Paracelsus first described, although some chemicals that act as hormone disruptors actually have a greater negative effect at lower concentrations. Either way, the idea of becoming completely toxic free is something of a pipe dream.

There are fundamental differences between toxics, toxicants and toxins. Toxics refer to external chemicals, including synthetic or artificial ones, which cause harm to humans and other living things or the environment. Toxicants are harmful synthetic chemicals that have been introduced to the environment (e.g. pesticides). Toxins are poisons produced by an animal (e.g. snake venom) or a plant (e.g. cyanide found in raw apricot kernels).

'Natural' is a buzzword commonly used on product packaging as part of tempting advertising slogans, but what does it really mean? It's easy to assume that natural chemicals are safe, and that those produced in a laboratory are riskier, but it ain't necessarily so. The distinction between natural and synthetic ingredients is far from black and white. One product might contain a mix of natural and synthetic constituents, and some chemicals are processed in order to mimic naturally occurring ones. Natural formulations aren't automatically less toxic than synthetic ones, or vice versa.

At first glance, the meaning of 'natural' or even 'naturally derived' might seem fairly simple, but it's important to look beyond the labelling and question what this implies. The degree to which any product is natural is twofold.

Firstly, it depends on whether the ingredients have been sourced from plants, minerals, marine resources or animals, or are petrochemicals derived from fossil fuels and are therefore synthetic. Secondly, it's important to consider how the ingredients have been processed or modified during production. So, everything lies somewhere on a spectrum and there are many varying degrees of 'natural'. So-called 'natural ingredients' can be toxic, yet aren't always regulated as stringently.

Naturally occurring ingredients are unprocessed and used in their natural state, such as raw honey or seaweed. Naturally derived, physically processed ingredients include raw, unrefined oils and butters that have been cold-pressed or filtered, though their molecular composition has remained the same. Other physical processes include the distillation used to produce essential oils or the extraction of certain plant-based ingredients. Naturally derived, chemically processed ingredients are naturally occurring substances that have undergone synthetic processing to structurally alter their composition. To make natural soap, for example, plant oils undergo a reaction with sodium hydroxide to form soap molecules.

Nature-identical ingredients are manufactured in a lab, but are chemically the same as those that occur in nature. Synthetic ingredients are created and processed in a lab, but bear absolutely no resemblance to anything originating in the natural world. Petrochemicals are synthetic chemicals made from fossil fuels (i.e. petroleum or natural gas), many of which have complicated scientific names that are hard to decipher. Petrochemicals are used to make everything from packaging and clothing to laundry detergents and fertilizer. Plastics are made from petrochemicals too and 98 per cent of single-use plastics are manufactured from fossil fuels.[1]

But before we decide whether or not natural products are healthier for us and better for the environment, we need to clarify a few things. The big picture is a melting pot of the sourcing, processing, application and disposal of the product.

Natural does not always mean safe and gentle. Some naturally occurring ingredients can have powerful effects on our bodies. Botanicals such as tea tree oil, eucalyptus and rosemary can trigger allergic reactions or skin irritation, and everyone has varying degrees of sensitivity too. Synthetic preservatives make a product less likely to spoil, whereas natural formulations without preservatives tend to have a shorter shelf life. In the US and EU, botanical and synthetic ingredients must meet the same regulatory requirements, regardless of the source, but in both cases, chemicals are designed to perform a particular function (like preserve a formulation) and their biological activity (such as hormone disruption) is often not evaluated before they are used in products.

Naturally occurring ingredients must be better for the environment than synthetics manufactured in a lab, though, surely? Advocates of natural believe that the regenerative or organic farming practices used to cultivate natural ingredients are supporting the ecosystems, improving soil health and protecting biodiversity. But, depending on how, when and where something is harvested, it might deplete the environment, just as making compost from peat is detrimental to peatland habitats.

Manufacturers of lab-produced synthetics will argue that their methods don't exhaust the planet of its resources and that they're able to produce substances on a larger, commercially viable scale without limitations of availability that might affect sourcing of naturally occurring ingredients.

Squalene, an oil found in the livers of sharks, is used in some vaccines and cosmetics. Around 3,000 sharks are required to extract a tonne of squalene, but to avoid threatening shark populations, scientists are testing a synthetic equivalent made from fermented sugar cane.

But while it is possible to make synthetic chemicals using renewable raw materials, most synthetic organic chemicals are derived from fossil fuels – predominantly petroleum and natural gas, and coal – using energy-intensive processes that can result in hazardous waste and toxic emissions. Between 1930 and 2000, global production of synthetic chemicals increased from 1 million to 400 million tonnes each year.[2] The chemicals industry is the production sector that uses the most energy in the world, resulting in vast greenhouse gas emissions which fuel the climate crisis.[3] So the use of petrochemicals is inextricably linked to the climate crisis. According to the International Energy Agency, demand is surging more now than ever before and petrochemicals are 'rapidly becoming the largest driver of global oil demand'.[4]

By volume, a huge 62 per cent of chemicals used in the EU are hazardous to human health and the environment.[5]

Every formulation is a compromise between price, effectiveness, aesthetics, performance and ethics. It's a pay-off and the environment often loses out. Environmental burden is not just about sourcing ingredients. What happens to a cleaning solution once it gets rinsed down the plughole or where does the content of the aerosol you spray around your home go? Disposal is key and when natural ingredients eventually return to the water cycle they'll be biodegradable, unlike many synthetic chemicals, some of which can produce serious and persistent pollution.

Some chemicals are more bioavailable or more easily absorbed than others. Medical drugs need to be readily bioavailable so that they can be taken up by the body and work effectively. Often, natural substances are absorbed faster by the blood than synthetic versions that aren't processed in the same way. The bioavailability of toxic chemicals is as relevant to human health as it is to that of the environment. In the soil, if toxic chemicals are bioavailable, uptake by plants and animals can cause detrimental effects. So, bioavailability needs to be considered when calculating the ongoing risk of a chemical once in the environment. That said, a lack of bioavailability shouldn't be used as an excuse to leave contaminants in the environment.

Just as public awareness about plastic pollution has grown exponentially in recent years, it's time now to consider what's in the furniture we have in our home, the creams we apply to nourish our skin, the clothes we wear and the sprays we use to keep everything clean. Since the world woke up to the horror of single-use materials, there has been a rise in recycled, recyclable and compostable packaging options. Now, the same logic must be applied to what's inside those bottles or how a fabric has been made, then you need to evaluate whether that fits in with your own ethos. If you eat a vegan diet, you might be surprised to find out that there's probably animal fat in your laundry detergent. If you buy organic food, perhaps think about whether the ingredients in your skincare products are organic too, and if you disagree with testing on animals, look for cruelty-free certification. By knowing your options, you'll be better able to make the choices that best suit you and your family.

Don't be duped

Greenwashing happens when a brand makes incorrect or confusing claims about environmental practices or products, often misleading consumers via outlandish marketing on labels, packaging or adverts that falsely imply that a product is sustainable in order to generate more sales. It was first coined as a term in 1986 by American environmentalist Jay Westerveld, who saw the hypocrisy in a hotel's campaign to encourage guests to reuse their towels to 'save the planet' while proper recycling was lacking throughout the hotel. He criticized the hotel for wanting to save money on electricity and laundry costs rather than saving resources. So, is a brand's PR spin in line with their actions behind the scenes?

Is that deception deliberate? As consumers, we want to do the right thing and we're more likely to buy products that say they are environmentally friendly. We want to be able to trust sustainable solutions. Equally, it's all too easy for companies to be tempted to fall into the greenwash trap and use 'natural' or 'eco' language that is permitted and is likely to improve sales, yet their assertions don't have to be verified.

The trick to spotting greenwashing is to be super vigilant. As we become increasingly aware of environmental issues and the impact that certain hazardous chemicals might have on our own health, it's crucial to look beyond the label and keep an eye out for greenwashing. If you know what to be wary of, it can be easy to spot when you're shopping and scanning the options available.

THE LOGOS YOU CAN TRUST

Certified claims should never be a substitute for robust laws, but these schemes are good indicators that certain proof points have been verified.

Certified organic: For food and cosmetics produced with minimal agricultural chemical input, look for the international Soil Association symbol and others including USDA Organic in the US, BioGro NZ in New Zealand and AROS or Asia Regional Organic Standard.

soilassociation.org

Indoor Air Controlled: From A+ for low emissions to C for high emissions, ratings show the toxicity risk from inhalation of products in indoor air.

air-label.com

Fairmined and Fairtrade: For responsibly mined gold and precious metals extracted with safe handling of chemicals and for food crops and textiles farmed without 200 hazardous pesticides.

fairmined.org, and *fairtrade.net*

Made in Green: This label by OEKO-TEX label indicates that textiles and leather have been tested for harmful substances and manufactured under socially responsible working conditions.

oeko-tex.com

Global Organic Textile Standard (GOTS): These textiles consist of at least 70 per cent natural fibres originating from organic farming practices and a maximum of 30 per cent synthetic fibres.

global-standard.org

Vegan: More than 45,000 products with the Vegan Society 'sunflower' trademark are free from animal ingredients and animal testing.

vegansociety.com

Cruelty Free International: The 'leaping bunny' logo is the gold standard for cosmetics and household cleaning products produced without animal testing.

crueltyfreeinternational.org

Green Seal: More than 33,000 household products, food packaging and construction materials made without toxic chemicals and processes have this certification.

greenseal.org

GreenScreen: GreenScreen Certified promotes the use of safer chemicals in the manufacturing of furniture, firefighting foam, fabrics and textiles.

greenscreenchemicals.org/certified

COSMOS Standard: For cosmetics made using responsibly sourced organic and natural ingredients plus environmentally friendly processing.

cosmos-standard.org

B Corp: For businesses that balance profit with purpose, this rigorous assessment considers every aspect of a company's impact, from social to environmental.

Bcorporation.net

Stay curious and always ask questions. If a claim sounds vague, it probably is. Ask yourself 'Where's the proof?' or 'What is the actual meaning?'. Look for definitions that are clear-cut and robust. So, if a label states that a certain cosmetic is toxin-free, your greenwash alarm bells should ring loud and clear.

Think about the whole supply chain, the ethics a company might follow, the environmental standards a product needs to adhere to or any independent audits they might undergo. A company might switch to certified organic cotton while still airfreighting goods around the world – sustainability requires a holistic approach and involves everything from sourcing to disposal.

Remember that a lot of marketing language involves offering solutions to customers. But what if the 'problem' really isn't an issue for you at all? Perhaps you don't need haircare products to straighten your curly locks or maybe you're quite happy with clean clothes that don't smell of lavender. Don't fall into the trap to fix things just for the sake of it. No product might be much better than a less toxic one.

Look for evidence of transparency. Scientific evidence has sometimes been concealed or manipulated to distort industry harms or skew evidence in favour of the chemical industry to maximize profit.[6] Nobody's perfect, so companies doing their best are usually honest about where they need to improve. Greenwashing stems from the fact that it is very difficult to define what 'green' actually means – there's no one gold standard. Claims are meaningless until they are proven and backed up by rigorous certification and, even so, a product might include only tiny amounts of one certified organic ingredient – so always read the label.

Lots of language remains unregulated. Many words or phrases like 'natural', 'nature-inspired' and 'eco-friendly' are not governed and therefore are pretty pointless in terms of the chemistry of a product. A brand doesn't need to meet any specific requirements in order to use these words in advertising and promotion, so it's important to read between the tag lines. Beware of the manufacturer's own symbols that display unverified or irrelevant claims such as 'dermatologist recommended' or '100 per cent natural'. Question any big claims – labels such as 'biodegradable', 'compostable' and 'sustainable' are not officially regulated yet. Most compostable materials are intended for an industrial composting process; it doesn't automatically mean it will degrade in your home compost heap.

Recyclable is regularly misused on labels. In theory, most things can be recycled, but the reality is that we can only recycle items if there is an established waste stream for it. That requires collecting, sorting, the right tech and a viable market for the recycled material. Plastic packaging alone is made from seven types of resin and a range of recycling symbols can be found, and some types such as PET (polyethylene terephthalate) are more recyclable than, say, polystyrene. But to add to the confusion, it's easy to be duped by the 'Green Dot' trademark, commonly found on supermarket products. The circular arrows imply recyclability, but in fact this European symbol just shows that the company has paid towards the cost of recycling.

There's a growing trend for non-toxic, pure products that are labelled as 'free from', 'with no nasties' or 'paraben-free'. But real transparency is about disclosing ingredients, so be wary when labels highlight what *isn't* being used to make something. Perhaps this free-from tendency follows on from fairly extreme interpretations of clean eating, but eliminating entire groups of nutritious wholegrains isn't particularly healthy and avoiding all processed foods isn't practical for most. Descriptions like these should all be flagged as warning signs.

Don't be deceived by gorgeous pictures, wholesome voiceovers or earthy colour schemes. Images of natural surroundings, beautiful landscapes and gentle floral palettes can give a false impression when ingredients might still have been manufactured in a lab and contain substances that are toxic to the environment. Natural fragrance is a misnomer too – they are synthetics made in a chemistry lab. Organic water raises another red flag. Water can't be certified as organic and some people consider it a filler, so there's an increasing number of waterless or anhydrous products, such as oil-based serums or face masks, that come in dry powder form. These can be 100 per cent organic and claim to be better for the environment because they help to conserve water and are lighter to transport. If a product states that its ingredients are, say, 80 per cent organic, check whether the remaining 20 per cent is water. The addition of aqua throws up another issue – the need to add preservatives or chemicals that stop microbes from growing in the water.

As consumers, we have the power to make big corporates and smaller brands more open about how they produce something – who is involved, which raw materials do they use and how were they cultivated or collected, who is manufacturing this and what process is used, how is it transported and what the ingredients might be. Take fossil fuels, for example. Mineral oil is a by-product of the petroleum industry used in Johnson's Baby Oil and Bio-Oil. Petroleum jelly used in Vaseline works by locking in moisture, acting like a barrier. Yes, they are deemed safe, but consumer choice comes down to personal preferences too, and general acceptance of these products will most likely shift as we move to a low-carbon society that relies far less on fossil fuels.

If you're still unsure about a product, you can always email or phone a company to ask for more information about its contents, query any doubtful claims and ask for evidence. Directly demanding this level of transparency sends a powerful message that, collectively, consumers are aware and really care. We need a shift in corporate consciousness, and we have the power to call out greenwashing and hold these retail brands to account. If you're really concerned, you can flag up misinformation to Trading Standards, the Advertising Standards Authority or the equivalent wherever you live.

Shoppers generally don't spend much time choosing between rival products, and so examples of misleading claims often focus on making key trigger words large, eye-catching and dominant, therefore directing attention away from other information about the ingredients. Your bubble bath might be 'inspired by nature' or some kitchen cleaners show the words '100 per cent natural' in huge type, but that might only actually relate to one ingredient. Vegan materials might just be plastic. It all comes down to legalities and these loopholes exist so consumer-friendly marketing terms persist, but alarm bells should start to ring when you see them.

Like the words within a stick of seaside rock, true sustainability runs through every aspect of a business and every part of a product, from eco-friendly packaging and fair labour to sustainably sourced ingredients that are processed in an ethical way without harmfully impacting the environment. It definitely shouldn't be an afterthought or a tick-box exercise. Integrity is about going that extra mile to do the right thing, even when it might be really difficult. Transparency is about showing what's not perfect too, and having a game plan with measurable targets to reach that goal. This also applies to us. There's no joy in aiming for perfection. That's impossible. But if we all do our bit, make informed choices that sit well with our own values and have a tangible positive impact, we can live a healthier lifestyle that won't cost the earth.

2

A Chemical Cocktail

From the moment we wake up in the morning and wash our face, moisturize, brush our teeth and pop on some deodorant, perfume or make-up, our everyday exposure to chemicals begins. A recent Californian study found that women used on average eight personal care products a day, with some using up to thirty on a daily basis. A huge 70 per cent of the women surveyed preferred scented products to unscented options, a choice that significantly increases the number of different chemical ingredients they interact with.[7] But the products we consciously use on our bodies are only a very small piece of the jigsaw puzzle. We're exposed to so many chemicals through the food we eat, the clothes we wear, the jobs we do and the buildings we live in.

To understand how this chemical cocktail affects us, we need to clarify a few things. Every item we own, from furniture and toys to shoes and saucepans, is just a snapshot of one of many long supply chains. There are some chemicals that might be most toxic during manufacture or once something gets thrown away. Some chemicals stay in our bodies, others accumulate in the environment, many do both.

Exposures add up. If chemicals aren't broken down or excreted, they can accumulate in our blood or get stored in fatty tissues. Some chemicals linger for a lifetime. Even a newborn baby has chemicals in its blood inherited during pregnancy from its mother.

What's your body burden?

Our body does its best to protect us. Our guts, lungs, skin and blood act as barriers, allowing certain things in but keeping others out. Stress, allergies, disease and exposure to toxic chemicals can all jeopardize the body's ability to function properly and prevent these barriers working at their optimum level. Certain individuals are more susceptible to particular toxic chemicals, and this varies with age or stage of life – everyone's threshold differs.

The toxicity or effect of a toxic chemical depends on the type of substance and on the dose; that's how much we're exposed to and how regularly that happens. One dose of benzene or mercury will have much more far-reaching impacts than one dose of methylparaben, a preservative added to cosmetics that gets broken up more easily in our bodies. But even chemicals that are essential to our survival can be deadly when consumed to excess. Chemicals vary in how long they take to break down in the body and for their potency to be reduced – some persist indefinitely, others get metabolized in just a few hours. And sometimes, not enough is known about the toxicity of chemicals before they are licensed for use. In 2019, the European Chemicals Agency found that important safety information is missing for 71 per cent of substances reviewed.[8]

Former Olympic ski champion Stine Lise Hattestad Bratsberg of Norway was one of the first people to take UNEP's body burden blood test. As a healthy athlete who didn't drink or smoke, she didn't expect many toxic chemicals to be detected and so was shocked by the results. Her test showed up chemicals from the ski wax she uses, as well as mercury – probably because she eats a lot of fish – plus high levels of a Japanese pesticide found only in traditional sleeping mats that she slept on while ski training in Japan. 'Now I eat organic food when I can, but that's a complex issue,' she explained to me. 'For some families, if they don't use fertilizer they have no food. As consumers, we need to ask more critical questions. Industries need to be more transparent and we have to take this seriously.'

Your body burden is influenced by the amount of toxic chemicals you are exposed to, or your toxic load. Your toxic load won't ever be zero and our bodies are designed to cope with this to some extent, but a high toxic load can cause problems. To a certain extent, we can abate our toxic load by changing various aspects of our lifestyle. By simplifying our beauty routines, household cleaning rituals and shopping habits, we can reduce our toxic load to a more manageable level with fewer ingredients at lower doses. By arming yourself with robust knowledge that's based on science, you can become a more empowered consumer. The first step is to simply pay more attention to the ingredients listed on product labels for everything you buy in your home, and consider the impacts of the manufacture and disposal of each of these things too.

Body burden tests aren't openly available to the general public, so when Dr Robin Dodson, an environmental exposure scientist at the Silent Spring Institute in Massachusetts, US, asked for people to sign up to a crowd-sourced biomonitoring project, she was inundated. People desperately wanted to find out more about the chemicals in their bodies. She tested urine samples from 726 people for ten common hormone-disrupting chemicals that tend to be cleared by the body within a couple of days. Having asked each participant to complete a detailed questionnaire, Dodson found that people who paid close attention to the ingredient labels on products and actively avoided specific hormone-disrupting chemicals had a lower overall chemical body burden.

With chemicals that aren't persistent, there's a huge upside. Intervention makes such a real and swift impact. 'With better chemical regulation, we can do something today to reduce the effect in three days' time,' Dodson told me. 'Many of these chemicals shouldn't even be in everyday products to begin with. I shouldn't need an advanced degree to buy products for my family; I should be able to trust that when I go to the store, I'm buying a product that doesn't contain chemicals that might affect my family.' But when Dodson had her own urine tested, she discovered that some chemicals are impossible to avoid, even with the best intentions: 'I had one of the highest levels of bisphenol S [one of the industrial chemicals used to make plastics and food packaging], but I have no idea where it came from,' she told me. 'I scrutinize all my labels and there are very few plastics in my house. I'm an empowered consumer, but I still can't avoid all these chemicals, even with a PhD in environmental health.'

The health impacts

Pollution is the world's biggest environmental cause of death and disease, and some contaminants, such as lead, cause harm to environments, people and animals – these are known as 'one health' issues. Around 9 million people die prematurely as a result of pollution every year.[9] But unlike infectious diseases, which tend to cause a quicker, short-term response, many of the chronic diseases caused by chemical pollution take years to develop, so the connection between cause and effect isn't as obvious. By understanding more about the negative effects that some chemicals have on our health, whether we experience them occasionally at a high concentration or consistently at low levels, we can work out how best to avoid the worst offenders.

Chemicals of concern can be toxic in many ways. Some act as carcinogens, endocrine disruptors or neurotoxicants. Some act as all three at once – chemicals called phthalates that are used to make plastics soft and as fixatives for fragrances, for example. Let's explore how chemicals affect the body in these different ways.

Carcinogens

Cancer-causing agents or carcinogens include viruses, hormones and toxic chemicals. Chemical carcinogens include synthetic and naturally occurring substances. Some mutate our genetic code or DNA; others may cause cells to divide more quickly and increase the chance of DNA mutations occurring. For many chemical carcinogens, effects arise after prolonged exposure, but carcinogens don't always cause cancer. They raise the risk of someone's chance of getting cancer. Many factors are at play, from the person's genetic predisposition to the length of exposure, so it's really difficult to test for cancer-causing effects. For instance, smoking is a known cause of cancer but not everyone who smokes will get lung cancer.

Endocrine disruptors

Endocrine disruptors are synthetic or natural chemicals that are made outside the body and interfere with the body's hormone system and affect our reproduction, growth and development, metabolism, immune system or behaviour. Endocrine disruptors might mimic a particular hormone – industrial chemicals containing chlorine can mimic thyroid hormones – and trigger a reaction or block the hormone from being absorbed completely and prevent its normal function. The rise in endocrine-disrupting chemicals, including some flame retardants and pesticides, has been linked to reduced IQ levels in children, attention deficit hyperactivity disorder and autism spectrum disorder.[10]

Depending on the timing of exposure, very low levels of endocrine disruptors can trigger long-lasting effects. Our hormone levels and functions change throughout our lifetime. For menstruating women, hormones fluctuate during every monthly cycle. For babies and children, certain amounts of hormones are required at just the right time. For pregnant women, thyroid hormone levels are crucial to the healthy brain development of the foetus, so they can be more prone to the effects of endocrine-disrupting chemicals, especially during the first trimester. Some of the phthalates, plasticizer chemicals that are used to make plastics and fragrances, are known to shorten the anogenital distance in baby boys – this is the gap between the anus and the base of the penis. The shorter the distance the greater the risk of problems later in life, such as low sperm count and male genital birth defects, following the mother's exposure to phthalates during the first trimester of pregnancy.[11]

But some standardized test protocols aren't designed to pick up the subtle effects that chemicals might have on the endocrine system. Major industrial chemical firms invest approximately 3 per cent of their chemical sales money in research and development[12] but chemical safety tests don't equate to real-life experience.

FEELING SENSITIVE?

In the case of sick building syndrome (the name given to symptoms you get when you're in a particular building), formaldehyde vapours, asbestos dust or synthetic fragrances occasionally trigger short-lived reactions such as headaches or dizziness while in a particular building. Whereas, in people with multiple chemical sensitivity, reactions to certain chemicals build up over time and skin irritation might begin after years of safely using a washing-up liquid or fragranced perfumes because someone's immune system gets overloaded. Dr Claudia Miller, an environmental health expert at the University of Texas Health Science Center, in the US, has developed a straightforward questionnaire called the Quick Environmental Exposure and Sensitivity Inventory (QEESI) to assess an individual's likelihood of developing chemical intolerance. She studied how people react to cleaning agents, solvents, volatile organic compounds (VOCs) or exhaust fumes, and theorized that a single major exposure, such as a chemical spillage, could cause a person to lose tolerance to synthetic chemicals that previously didn't bother them. Subsequent low-level chemical exposures then trigger symptoms such as digestive problems, trouble concentrating, muscle pain and fatigue. She calls this toxicant-induced loss of tolerance (TILT).

Neurotoxicants

Synthetic chemicals that disrupt communication in the brain and nervous system are known as neurotoxicants; naturally occurring ones that do this are neurotoxins. Alcohol and heroin have neurotoxic effects, as do industrial chemicals including mercury, arsenic, organophosphate pesticides and some flame retardants. Right from the moment of conception, neurotoxic chemicals influence the formation of a baby's brain, its development and subsequent behaviour, and IQ.

A case in point is lead. While this naturally occurring metal does cause long-term harm in adults, young children are particularly vulnerable, as first discovered during the 1970s by Professor Philip Landrigan, a paediatrician and epidemiologist from the Mount Sinai Children's Environmental Health Center in New York City. Once in the body, lead is distributed to the brain, bones, liver and kidneys, and can result in an increase in blood pressure. Lead poisoning can cause premature death, and in pregnant women it can bring about stillbirth and miscarriage. According to an investigation by the UN's children's charity UNICEF, the most widespread damage occurs in children who are particularly susceptible to disrupted brain development and absorb up to five times as much ingested lead as adults. One in three children – up to 800 million worldwide – have dangerously high blood lead levels.[13] Lead is associated with lowered IQ, behavioural difficulties and learning problems. There is no safe level of exposure to lead, the effects are irreversible, yet lead exposure is completely preventable if we put a stop to pollution at source.

Chemicals are not child's play

The health effects of toxic chemicals can be amplified when exposure happens during particular key stages or 'critical windows' of development. In an adult, the body's organs and systems are fully formed. For foetuses, babies and children, chemical exposure could interfere, alter or disrupt processes that only happen during these phases of development, just as air pollution poses a greater risk to under-eighteens because their lungs are still growing. Other factors are also at play. Children are potentially much more vulnerable to chemical exposure than adults because their metabolism differs. Furthermore, children have a greater proportion of their life left to live, so in theory they have longer to accumulate problematic chemicals inside their bodies too.

Children are not simply small adults. Their behaviour is vastly different, as are their diets and their use of consumer products. Of course, we have childproof caps on medicine bottles and perhaps safety latches on the cleaning cupboard where the bleach lives. But children are curious creatures. They like to get their hands on things and they often use their mouths to test things out. They might suck on their sleeves, eat a handful of soil that they've dug up in the garden, pour an excessive amount of bubble bath into their bath or lick spilt food straight off the dusty floor. Yes, really.

They might not use make-up or clean the oven, but they're more tactile, very active and their sweat could increase the rate of chemical absorption through the open pores in the skin. The clothes we buy for children might well be more about function than fashion and therefore more likely to contain added chemicals to make them waterproof, stain-resistant or wrinkle-free. Children have a higher body-surface-area-to-weight ratio than adults, so their external surfaces – their skin, lungs and gut – could potentially absorb more toxic substances. Relative to their body weight, they eat more food, drink more water and breathe more air than adults. So when experts assess potential risks of toxic chemicals, it's not simply a case of extrapolating impacts of exposure from adults to

A PUNNET A DAY

Toxic chemicals put extra strain on the body's natural detox mechanisms – mainly the liver and kidneys, plus the gut, skin and lungs. Exposure to chemical pollutants can cause inflammation in the body and the production of excessive free radicals – these are unstable molecules that can trigger a process called oxidative stress. This damages the DNA inside the body's cells and can lead to degenerative diseases. Free radicals are produced naturally in the body, but smoking, drinking alcohol, regularly eating fried foods and toxic chemicals all contribute to free-radical production – as such, these chemicals are known as pro-oxidants. Antioxidants, sometimes known as free-radical scavengers, help to neutralize some free radicals, therefore helping to protect cells from damage. It makes sense that the more alcohol you consume and the more you smoke, the harder your liver will be working before any other toxic chemicals are even added into the mix. We can all make a conscious effort to eat more antioxidant-rich foods every day. Things like green, leafy veg such as spinach, lots of nuts and organic blueberries, raspberries or strawberries by the punnet.

children according to their size; younger people might well be more susceptible to toxic chemicals.[14] It's far from linear.

My top ten toxics

Which toxic chemicals should we be most worried about? Unfortunately, there's no simple answer. Some groups of chemicals result in the greatest deaths or severely damage the environment in certain hot spots, some are noticeably more prevalent in our homes but far from essential, while others are used without regulation before the consequences are understood. By limiting our use of the following ten groups of problematic toxics, we can all play our part in reducing chemical pollution.

- PFASs
- Pesticides
- Flame retardants
- Metals
- Plasticizers
- Antimicrobials
- Some solvents
- Particulates (PM2.5)
- Nanos
- Unflushables

PFASs

What are they?

This diverse family of synthetic chemicals includes more than 9,000 substances that repel dirt, food residues, water or greasy stains, so PFASs (perfluoroalkyl and polyfluoroalkyl substances) are used to make products non-stick, waterproof, wrinkle-proof or stain-resistant. First introduced in the 1940s, PFASs are extremely stable chemicals, so they are often referred to as 'forever chemicals'. They are found in clothing, cookware, cosmetics, carpets, building materials, food packaging, firefighting foam and many other household items. There are no natural sources of PFASs, but they are becoming increasingly ubiquitous.

What's the problem?

Because PFASs don't break down easily, they persist in the environment and can build up in our bodies and in nature. PFAS production can pose a huge hazard to workers and these chemicals can escape from factories, industrial sites and military bases where they are used in firefighting foams. PFASs end up in the soil, atmosphere, oceans, crops, drinking water, breast milk and our blood. PFASs that have already been phased out still end up in food grown in contaminated areas while exposure within the general population is more likely to occur via the direct use of personal care products. Some PFAS chemicals are thought to break off PFAS-treated everyday products such as carpets and clothing, and end up polluting indoor air and accumulating in dust, so they can potentially be inhaled.

Many PFAS chemicals are endocrine disruptors and possible carcinogens, but only a couple are globally regulated, including PFOS (perfluorooctane sulfonate) that was used to make Gore-Tex and Scotchgard water-repellent. Despite restrictions, older consumer products may still contain PFOS. Teflon has been made with a PFAS chemical called C8 or PFOA (perfluorooctanoic acid) that's now banned due to links with six diseases in people exposed to high levels during manufacturing, including kidney and testicular cancers, decreased fertility, elevated cholesterol and thyroid dysfunction. These banned PFASs have been replaced with chemical cousins that are potentially just as harmful. While wearing a Gore-Tex jacket or cooking with Teflon pans won't kill you, their manufacture and disposal releases persistent forever chemicals into the environment.

Pesticides

What are they?

Pesticides are toxic by design. This group of chemicals includes insecticides that kill bugs, fungicides that kill fungi and herbicides that kill weeds. They are used widely in industrial agriculture as well as in our homes and gardens. In addition to the chemicals available for use on our lawns and vegetable patches, flea treatments for pets, mosquito deterrents, mothballs and headlice shampoo all contain pesticides too. Residues can be found in non-organic cotton textiles and remain on certain fruit and veg. Imported foods sometimes contain traces of banned pesticides as regulations aren't consistent globally.

What's the problem?

Certain highly hazardous pesticides cause extreme harm to human health during use, they pose a serious risk if accidentally ingested, and when it rains, any excess runs off into the surrounding environment where they can be toxic to wildlife. In an ideal world, these most dangerous of pesticides would be banned and farmers would all switch to safer alternatives and non-chemical pest control.

Flame retardants

What are they?

Hundreds of different chemicals have been added to manufactured materials since the 1970s to supposedly make them less flammable or delay the production of flames or the spread of fire. Flame retardants can be found in household furnishings, electronics and construction materials, as well as in the plastic components and fabrics used to make cars, trains and planes. They are often categorized by their chemical properties, and groups include organophosphate flame retardants and halogenated, particularly brominated, flame retardants.

What's the problem?

The fire-safety benefits are debatable and the risks far outweigh any advantages. Flame retardants known as polybrominated diphenyl ethers (PBDEs) are endocrine disruptors and have been detected in human blood, breast milk and newborn babies. PBDEs are being phased out, but they continue to endure and are often replaced with persistent organophosphate flame retardants, which also cause health harm. After manufacturing, flame retardants can be released into the air or 'off-gassed' from furniture, electrical equipment, foam mattresses, building insulation and textiles. Then they accumulate in house dust and in the food chain. Once disposed of, they keep on contaminating the environment.

Metals

What are they?

Heavy metals – metal substances that have relatively high density – exist naturally in the Earth's crust, but human activity, extraction and industrial processes such as mining or burning fossil fuels release metals into the soil, air and water with huge implications. The World Health Organization (WHO) is most concerned about lead, mercury, arsenic and cadmium.

What's the problem?

Once particles begin to circulate in the air, they can be breathed in, land on crops and enter the food chain, or wash into the waterways and build up in wildlife. People whose jobs expose them to these metals are at greatest risk, but the effects of contamination are widespread globally. Mercury still gets released during artisanal goldmining, lead leaches out during informal recycling of car batteries, shiny cadmium is sometimes added to cheap costume jewellery and arsenic is used to make some pesticides and wood preservatives. At the moment, organic certification doesn't test for the presence of heavy metals in food. Once the environment has been contaminated by heavy metals, it's expensive and challenging to detect and remove so the ideal solution involves switching to alternative processes that won't release these pollutants in the first place.

Plasticizers

What are they?

Plastics are made from many thousands of different chemicals, including phthalates and bisphenols. Phthalates are added to plastics to make them soft, strong and flexible, and they're used as fixatives for fragrances in personal care products. Forty or so types of bisphenols are utilized as hardening agents in the manufacture of most plastics, from baby bottles and teething toys; in tin cans they form a barrier between the food inside and the can material; and with the thermal paper used to make till receipts, bisphenols act to develop the printed image when heat is applied.

What's the problem?

A quarter of the 10,500 compounds used to make plastic are substances of concern.[15] Plasticizers can leach from plastics to food and some phthalates and bisphenols are endocrine disruptors. Certain phthalates (with the catchy names DEHP, BBP, DBP and DIBP) have been banned in baby products in some countries because they can worsen asthma and allergies, as well as trigger early onset of puberty or cause fertility problems, but they remain in other household items that children use every day. Bisphenols can migrate out of packaging into food and drink, and can end up in house dust or even birds' eggs and sand on the beach. BPA (bisphenol A) mimics oestrogen and can affect male and female infertility and lead to breast and prostate cancers, among other conditions. BPA gets broken down by the liver and can be found in most people's urine, but many people are exposed on a daily basis, and even a small amount can have an endocrine-disrupting effect. It has been banned in some parts of the world, but is often replaced with similar alternatives such as BPS (bisphenol S) and BPF (bisphenol F), and the worry is that these could disrupt hormones in a similar way.

Antimicrobials

What are they?

Antimicrobial chemicals are added to products to kill bacteria and viruses, sometimes as disinfectants or preservatives. They can be beneficial because they can lengthen shelf life and stop cosmetics from going mouldy, and importantly they can be used to sterilize things in medical settings. For many years, triclosan has been added to liquid soaps and body washes, and triclocarban has been an ingredient within solid bars of soap. Other common antimicrobials are known as quats or quaternary ammonium compounds, and these potent disinfectants can be found in sprays, wipes and many household cleaners.

What's the problem?

Thousands of consumer products are made with antimicrobials – everything from soaps to yoga mats, kitchen worktops, clothing and food storage containers that claim to have anti-odour, antibacterial or antimicrobial qualities. But the overuse of antimicrobials for non-essential use is a big issue. When antimicrobials wash down the plughole into waterways, they contribute to antimicrobial resistance, whereby superbugs evolve that are harder to treat with antibiotics. Triclosan and triclocarban have been banned for certain uses such as liquid antibacterial soaps in the US and restricted in Europe and Japan, but are sometimes replaced with quats that can disrupt hormones and exacerbate allergic conditions.

Some solvents

What are they?

Solvents are used to dissolve or disperse other substances, but a few of them can have serious knock-on effects. Some are neurotoxicants, others are carcinogenic. Solvents are used in industrial manufacturing and they're found in oil-based paints, paint strippers, adhesives, wood finishes, shoe polish, some cosmetics and household cleaners. Some of the most toxic are benzene (found in some paints, glues and some air fresheners, plus car exhaust fumes), toluene (also known as methylbenzene and used in paints, glues and nail-varnish removers), perchloroethylene or PERC (used in traditional dry cleaning), chloroform (a strong-smelling liquid produced on a large scale as a precursor to Teflon, various refrigerants and pesticides) and formaldehyde, one of the best known VOCs (chemicals that evaporate easily at room temperature) and an airborne pollutant that acts as an irritant.

What's the problem?

Some solvents, such as benzene, are carcinogens; others like toluene are neurotoxicants. If people are exposed to high doses or over a long period of time, the health effects can accumulate, so working in certain industries can be hazardous. For example, formaldehyde is found in building insulation, resins used to make plywood or composite furniture, some paints and floor lacquers. New foam mattresses, upholstery and chemical hair-straightening products can all off-gas fumes of formaldehyde. Formaldehyde is produced and broken down by the body in tiny amounts, but high doses inhaled over long periods of time are associated with skin allergies, respiratory issues and cancer. The production, use and disposal of toxic solvents can all be harmful and not only to people but to the environment too. Once released into the atmosphere, organic solvents react in sunlight to produce an air pollutant called ground-level ozone, which is damaging to forests, crops and wildlife at high concentrations.

Particulates (PM2.5)

What are they?

Particulate matter forms an aerosol, comprising a complex mix of liquid droplets and solid particles such as soot, dust, tyre rubber and brake wear from vehicles, smoke and metals. Particles or particulate matter (PM) are classified by size and are key indicators of air pollution. Coarse PM10, like dust or pollen, are 10 micrometres or smaller. PM2.5 are 2.5 micrometres or smaller and include invisible particles produced by traffic fumes, log burners, conventional ovens and even toy cars with electric motors.

What's the problem?

Once inhaled, PM2.5, including metals and nano-sized particles, can travel deep into the lungs and bloodstreams of people and other animals. PM2.5 is the fifth leading risk factor for fatalities, accounting for 4.2 million deaths according to the Global Burden of Disease estimates. WHO attributes 3.8 million deaths to indoor air pollution from fires for cooking and heating. In the UK, domestic combustion (mainly from open fires and wood-burning stoves) produces 45 per cent of PM2.5 – that's way more than road traffic, which accounts for just 12 per cent.[16]

Particulate matter can travel long distances in the air and damage crops and forests, or change the nutrient balance in rivers and seas. Haze is formed when sunlight hits areas of concentrated particulates emitted from industrial facilities and traffic, making the atmosphere murky and less transparent. Over time, this can prevent plants from photosynthesizing and growing properly, and reduce the activity of insect pollinators that are vital for healthy ecosystems and food crops. So, reducing air pollution, both indoors and outside, is key.

Nanos

What are they?

'Nano' means 'dwarf' in Greek. Nanoparticles are smaller than 100 nanometres (nm) in size. That's super tiny – the average human hair is 80,000 nm wide. While nanoparticles exist in nature, and some manufactured ones have incredible applications, such as really specific drug delivery, the use of nanoparticles isn't yet well regulated. They can be added to all sorts of things – self-cleaning glass windows made with titanium dioxide or antibacterial gym leggings with nanosilver are just two examples.

What's the problem?

Labelling of nanos used in products isn't yet consistent globally and the addition of nanos is sometimes used as a marketing gimmick without considering the knock-on effects. So much is unknown about how nanos behave inside our bodies and what happens once they get released into the environment. The discharge of nanos isn't controlled, so they end up in the soil, air and water where they are impossible to retrieve. Nanosilver and titanium dioxide nanos are known to cause harm to some insects, and when used in pesticides, they could end up as toxic residues in our food. Silver can act as an important line of defence against microbes but resistance can develop, and some silver-tolerant soil bacteria have already been found.[17]

Health harm depends on how someone is exposed. Inhalation poses the biggest risk. In 2021, face masks coated with a nanomaterial called graphene that claimed to be antimicrobial were banned in Canada after concerns that nanos that have been linked to lung cancer would be directly inhaled. Inside our bodies, some nanos might transfer to babies via the placenta (some do so in pregnant rats[18]) and because they're so small, they could potentially be more easily absorbed into the bloodstream or into cells where they could damage DNA.

Unflushables

What are they?

So much more than just poo, pee and paper gets flushed down the loo, and way too many chemicals are washed down the plughole that shouldn't be. Everything that enters the water system, from plastic tampon applicators and cotton buds to toilet cleaners and sink unblockers or fats, oils and grease (FOG), ends up somewhere.

What's the problem?

Once washed away, chemicals do get diluted, but it all adds up and some can really be quite costly to detox at the water-treatment stage. Many cosmetics and cleaning products contain hidden ingredients such as liquid polymers which, just like microplastics, are not biodegradable. Of equal concern is that the production of bleach, which is used as a disinfectant, is incredibly toxic and there's no need for such a harmful cleaning product in a residential setting. Take a look at the back of the bottle and you'll see a warning label featuring a dead fish, as bleach is so harmful to aquatic wildlife once washed away too.

Small items sneak through water treatment filters so it's not surprising that cotton-bud sticks are the third most common item found on European seashores.[19] Sewers often get blocked by plastic waste and FOG that solidifies further down the drain block. Water companies often run 'think before you flush' campaigns, but solutions lie much further upstream – tampon applicators and wet wipes should, at the very least, be made from materials that can biodegrade in the water system, or be banned altogether.

What's my body burden?

I had my blood tested for one hundred POPs by a laboratory in Norway with some surprising results. Professor van Bavel explained that my blood contained traces of chemicals that have been regulated and taken off the market decades ago like PCBs and DDE, a metabolite of the banned pesticide DDT. 'Levels are low, perhaps 5 per cent of the levels we saw in the 1980s, but it shows that once chemicals are in our society it's really hard to get rid of them and some chemicals will stay in your body throughout your lifetime,' he said. The test showed higher levels of oxychlordane, a chemical that comes from an insecticide called chlordane that's generally found at lower concentrations than DDT in the UK. Chlordane was banned in the EU in 1981. I was born a year before that and lived in the countryside, plus my mother came from a farming family, so I could have been exposed via the placenta and her breast milk. 'For some of the worst POPs, the half-life can be between two and thirty years – that's the time it takes for the concentration of that chemical in your body to halve,' explained van Bavel.

While banned brominated flame retardants were below the detection limits, the most concerning were PFASs. My blood test showed higher levels of PFOS and PFNA (perfluorononanoate) so van Bavel wondered whether I lived near a source of contamination, such as an airfield where fire-fighting foams are used frequently. I grew up nine miles from a busy airport, although PFASs are present in thousands of everyday products and PFNA has been used as a replacement since the phase-out of PFOS and PFOA so it's impossible to pin down one source.

According to van Bavel, my results were fairly average. 'We're all exposed to these kinds of chemicals and your levels are at an acceptable level from a human health perspective but if we didn't have any measures in place, those levels would rise and we'd see different toxicological effects.' That's why we all need to support better legislation and avoid switching to alternatives that could be equally as toxic.

By thinking of chemicals in terms of classes, and by sidestepping entire groups of these ten toxic ones unless their use is absolutely essential, we can massively reduce our personal toxic load and put less pollution pressure on the environment. While our own toxic load won't ever reach zero, streamlining the number of products used on a daily basis helps to minimize that chemical cocktail effect. Rather than getting hung up on specific names of chemicals, think broadly about the types of chemicals that products contain and aim to avoid the most hazardous ones. By making sure everything is as safe and healthy as possible for the most vulnerable members of our family, often babies and children, we can dramatically reduce our exposure and subsequent body burden. And even if some toxic chemicals aren't present in our homes right now, it's important to remember that our usage is just one phase of the entire life cycle of a product. Larger problems might be occurring upstream at the factory stage or downstream once we wash them away into the big, wide world.

3

Pathways to Pollution

Pollution is absolutely everywhere. There are no borders and it's moving around all the time. From the top of Mount Everest to the Arctic Ocean and the bottom of the 11 km-deep Mariana Trench in the Pacific, pollutants spread far from their source, contaminating the planet on a global scale. Some pollution happens naturally – caused by wildfires and volcanoes, for example – but we produce so much more in addition to that. Chemicals can get washed out from factories or from our homes into nearby waterways, leaching into streams, rivers, lakes and then flowing into the sea which acts as a sink for the pollution we create. Once released into the air, chemical droplets and particles, including heavy metals and microplastics, get transported by the wind and rain. When they land and are deposited on the soil, they can be absorbed by plants and crops, eventually ending up in the food we eat.

Most pollution occurs as a result of human activity disrupting the planet's natural balance. From mining to manufacturing, a lot of industries produce chemical emissions or effluent – that's the liquid waste that flows out of factories that sometimes reaches our rivers and seas. When heavy metals that exist naturally in the Earth's crust are released, or effluent from chemical manufacturing escapes, pollutants end up in far-flung places. More toxic chemicals get added to the mix when farmland is irrigated with polluted water, fertilizers, sewage or industrial sludge. Whether this chemical run-off comes from farms, factories, our vegetable patch or our kitchen sink, its actual toxicity comes down to the volume of what is released into the environment. Even organic, biodegradable products cause harm at high concentrations. And the household chemicals we use every day contribute to this environmental load whenever we wash something down the plughole, throw something away or spray something into the air inside our homes.

From the chemicals used to make our material goods to the chemicals that end up in the environment, the environmental footprint is much bigger than the snapshot we experience in our homes. Some pollutants persist indefinitely. The climate crisis also increases the risk of chemical pollution at historic landfill sites, especially at coastal locations. With increasing coastal erosion, flooding and rising sea levels, there's a greater chance that another wave of contaminants will get released into the ocean. It's another reason that sea defences need to be maintained in order to restrict pollutants to landfills where they can be managed properly.

A toxic legacy

Despite bans being introduced, some chemicals don't disappear or degrade even long after their production has ceased. A century ago, and even just fifty years ago, nobody realized that these legacy contaminants or forever chemicals would have such long-term consequences for people, wildlife, food chains and entire ecosystems.

Back in the 1920s, way before they were ever deemed hazardous, a family of 209 synthetic industrial chemicals called polychlorinated biphenyls (PCBs) were introduced, mainly for use as cooling fluids in machinery, in electrical goods, as flame retardants, paints and building sealants. They were super useful because they were stable, inflammable and heat resistant. Production peaked in the 1960s, but then it was discovered that PCBs affect the immune system of animals and people, act as endocrine disruptors and reduce fertility; plus they're classified as a probable carcinogen. Today, most PCBs are found in sealed landfill sites, old electrical equipment and prefabricated buildings made before the 1980s. That stability, once so invaluable, translates into persistence and ongoing pollution problems.

Unless you live right by an industrial or agricultural site, your most likely source of PCBs is food. PCBs continue to build up or bioaccumulate in fatty tissues of animals on land and in the sea, and once in the food chain, they can biomagnify – this happens when chemicals build up and reach higher concentrations in top predators than in their prey. That's why it can be riskier to regularly eat large fish such as tuna and swordfish than, say, mackerel, anchovies and sardines.

PCBs are just one type of persistent organic pollutant, or POP, that stay in air, water, soil, human blood and even breast milk; they're everywhere, despite being banned in the US, UK and Europe for about the last forty years. A list of twelve POPs known as the 'dirty dozen' – including PCBs and the synthetic insecticide DDT – were banned in 2004 by an international treaty called the Stockholm Convention on Persistent Organic Pollutants and more chemicals have since been added. To date, 184 parties have signed the treaty and promised to implement control measures, but putting this legislation into practice can be a very slow process. Some chemicals must be phased out entirely and eliminated, while for others, production is restricted. So even though the insecticide DDT can't be manufactured or applied as an insecticide in agriculture in most countries, it can still be used to control malaria, although the provision of malaria nets would be much less problematic. Existing stocks of chemicals need to be dealt with and disposed of appropriately too, but that's not easy either. The most effective way to get rid of PCBs is by incineration at extremely high temperatures, but this is expensive and sometimes logistically impossible. So even with widespread bans, pollution continues to spread.

Chemical pollution is fuelled by over-consumption, so if we buy less stuff, be more energy-efficient, avoid wasting food and travel more sustainably, we can drastically reduce our impacts.

Once released into the environment, POPs just keep moving around the planet. In hot, tropical climates near the equator, persistent and volatile pollutants can evaporate out of the soil into the air, then travel in air currents to cooler places. Here they might condense out in the rain and fall back onto the soil or ocean. This long-range transport is repeated again and again, moving chemicals vast distances in a series of seasonal hops towards the poles of the Earth. As temperatures rise with global heating, POPs will volatize even more easily into the atmosphere, so in the future, this 'grasshopper effect' could have an even greater impact on people, animals and places thousands of miles away, long after bans have been put in place.

This long-distance transport phenomenon is already causing a health crisis. In the North Atlantic, women living in the remote Faroe Islands have very high concentrations of flame retardants known as PBDEs (polybrominated diphenyl ethers) in their breast milk. These PBDEs act as a neurotoxicant, affecting brain development and reducing IQ in babies and children. These women live nowhere near industry or chemical manufacturing, and PBDEs have been phased out in many countries in recent decades. So how did this happen?

POPs like these get into the food chain and can affect communities far, far away from the source of contamination. In the Faroe Islands, it's traditional for people to eat a lot of seafood and hunt pilot whales for their meat. At the top of the food chain, pilot whales carry high levels of toxic chemicals, such as POPs and mercury, that have built up from all the prey they have eaten. Meanwhile, Inuit people living in the Canadian Arctic often have higher metal and POP body burdens in their blood than that of the general population of Canada, predominantly due to their diet of fish and marine mammals such as walrus and narwhal.[20]

CSI of the sea

Let's take a step back and consider what happens when these POPs accumulate in marine animals – particularly cetaceans (whales and dolphins) – that are high up the food chain and normally long-lived with a high proportion of body fat or blubber.

When a twenty-year-old female orca known as Lulu was found washed up dead after getting entangled in fishing gear on the Isle of Tiree, Scotland, in 2016, scientists claimed that she was 'one of the most contaminated individuals' they had ever studied. A post-mortem by the Scottish Marine Animal Stranding Scheme found that her blubber contained shockingly high levels of PCBs – more than eighty times the level at which they are known to affect the health of marine mammals – and her ovaries showed that she had never produced a calf. Lulu was part of the British Isles' only resident inshore population known as the West Coast Community, a pod that has been monitored since the 1990s. Out of ten adult orcas in this group, two are known to have died (including Lulu) and only two of the remaining eight have been sighted in recent years. No calves have been spotted with any of these orcas since photo identification surveys began. So could toxic chemicals be playing a part in this?

'As strandings scientists, we're generalists looking at lots of different threats,' Rob Deaville, project manager for the UK Cetacean Strandings Investigation Programme based at the Zoological Society of London, explained to me. Every year, he carries out 150 post-mortems on stranded harbour porpoises, dolphins and whales, assessing the health, age and possible causes of death. His studies include measuring various toxic chemicals in the subcutaneous fat or blubber of cetaceans.

A SAILOR'S PERSPECTIVE

Ocean advocate Emily Penn heads up eXXpedition, a series of all-women voyages around the world to research microplastics and toxic pollution. She had her blood tested for thirty-five chemicals banned by the UN that are known to be toxic to people: twenty-nine of them were present in her bloodstream! Keen to find out more, she organized body burden tests for twenty of her female crew members. 'One Californian had very high levels of flame retardant inside her – interestingly, this state has one of the world's highest levels of flame retardant usage. Another woman had high levels of mercury and had recently spent six months living in the Amazon eating fish downstream of a mine that uses a lot of mercury.' One sixty-seven-year-old crew member registered an 'off-the-chart' quantity of the pesticide DDT – she was the only one to have been alive before the DDT ban came into place in 1972. The DDT had been in her body since she was a girl.

But animals don't experience individual contaminants in isolation so it's meaningless to focus on just PCBs.

So Lulu's abnormally high levels of PCBs could have made her more susceptible to other factors and increased the risk of death. Deaville's detective work acts as a canary in the mine: 'We do post-mortems on stranded animals to gain insight into their pollutant levels and that might give a broader picture of what's happening at the ecosystem level.' Over the past three decades, post-mortems have analysed a whole range of pollutants, from DDT and brominated flame retardants to the antifouling paints used on boats, plus heavy metals such as lead and mercury. PFASs are currently an issue of concern and Deaville expects other troubling chemicals to be flagged up in the future.

In people, the effects of exposure to legacy contaminants like DDT pass down to future generations too, from women to their daughters and granddaughters. A groundbreaking study spanning three generations over sixty years shows that young women whose grandmothers were exposed to DDT in the 1960s before or during pregnancy have higher rates of obesity and earlier first menstrual periods, both established risk factors for breast cancer.[21] This is a powerful warning that ancestral exposure to forever chemicals like DDT continues to have an effect today.

But this shouldn't come as any surprise. Back in 1962, the American marine biologist Rachel Carson warned of the detrimental effects of agricultural pesticides such as DDT in her book *Silent Spring*. While her prophecy caused a huge sea change in opinion at the time, the problem of chemical residues hasn't gone away. If anything, it has got much worse. Aside from DDT, so many other potent chemicals are being sprayed on our crops today and leaching into the environment. Carson's alarm call about the poisonous legacy of pesticides remarkably still holds true, sixty years on.

I wonder whether we'll ever learn from our mistakes and treat emerging pollutants with the necessary caution? Some of the chemicals in use around the world today could become the legacy contaminants of the future. Scientists are only just beginning to understand some of the pathways of chemical pollution; in years to come, there may well be more we haven't yet thought of. We must plan ahead and take the worst chemicals out of production now to avoid extremely

where it's treated. In countries where that system is broken or sanitation infrastructure is non-existent, raw sewage becomes a pollutant. But even within nations that have good treatment systems, untreated sewage can be discharged straight into the environment through combined sewer overflows (CSOs). During storm surges when sewers get overloaded, water companies can discharge raw untreated human sewage and wastewater directly into river systems via pipes designed to be used only under exceptional circumstances. In 2019, in England alone, this happened more than 200,000 times.[25]

'Rivers should be the blue arteries and the lifeblood of communities, towns, cities, villages, but they are effectively being treated as open sewers,' warned Hugo Tagholm, CEO of environmental charity Surfers Against Sewage (SAS). 'We know that our sewage system is discharging a toxic mix of chemicals, antibiotic-resistant microbes and pathogens. It's our job to point out the problem, but the water companies, agriculture and industry have to work out how to stop this.' Despite this, Tagholm remains optimistic that legislation can be improved. 'With enough grass-roots rebellion, combined with the forced pressures going down, the government and industry won't be able to get away with this stuff for much longer.'

In places like France and Germany, swimming is commonplace at hundreds of designated inland bathing waters. England has zero. All 4,600 rivers, lakes and streams are polluted beyond legal limits and freshwater wildlife is declining as a result of sewage contamination. Eventually, that pollution reaches the ocean. In collaboration with the European Centre for Environment & Human Health, SAS ran a Beach Bum survey in 2018 whereby 300 people, half of them surfers, took a rectal swab for analysis by scientists.[26]

The survey found that surfers tended to swallow up to ten times more water than sea swimmers and were at higher risk of exposure to antibiotic-resistant E. coli bacteria present in coastal bathing waters.

Antibiotic resistance could make one of the pillars of modern medicine obsolete, but recreational use of coastal waters is just one route of exposure. Increased hygiene during Covid-19 means more antimicrobial soaps and disinfectants have been getting washed down the plughole into wastewater treatment plants, potentially leading to more antimicrobial resistance. We can be exposed through direct contact with other people and international travel, but the biggest contributor is intensive agriculture and global trade of the food we eat.

According to the UN Environmental Programme (UNEP), across the world 70 per cent of antibiotics are used by animals and up to 80 per cent of consumed antibiotics are excreted as trace residues in urine and faeces. Animal manure containing antibiotic residues and antibiotic-resistant bacteria might be spread on fields as fertilizer, potentially contaminating crops and water supplies. But so much of this comes down to antibiotic misuse. In countries such as the US, Australia, New Zealand and Canada, antibiotics are used as a growth promoter to make livestock grow faster. This practice was banned in the EU in 2006, but antibiotics are still widely used to prevent disease rather than treat it among poultry, farmed salmon and most of all livestock.

Antimicrobial resistance can't be eliminated, but it does need to be better managed. Removing chemical pollutants to make water safe enough to drink is expensive, but polluters must take financial and environmental responsibility for remediation. Microbiologists at the University of Exeter, UK, have developed a quick and easy test that regulators can use to determine safe levels of antibiotics before the release of treated sewage into waterways. Investment in systems that better cope with storm waters would be a major step towards improving the health of river systems. But before it even enters the water system, we need to make sure antibiotics are only ever used when absolutely necessary in people and animals.

The problem with pesticides

Since the 1970s, chemical agriculture has become big business. The world's five largest pesticide manufacturers make billions – a third of their income – from selling chemicals that are highly hazardous to people, animals or the environment. An investigation by Public Eye and Greenpeace's Unearthed found that Brazil – a country that is home to up to 20 per cent of the world's remaining biodiversity according to UNEP – buys more pesticides than any other country ($3,330 million), half of which includes highly hazardous varieties. Some of the most dangerous pesticides are banned or not approved for use in the EU, yet they can still be manufactured here and exported to other, often poorer, nations. Low-to-middle-income countries might have less choice about which agricultural chemicals they can afford or be permitted to use, and so there's huge inequity.

Pesticide use has huge ripple effects. In Japan, neonicotinoid pesticides or neonics are applied to the rice paddy fields and get washed off into nearby waterways. At Lake Shinji, this has caused aquatic insect and plankton numbers to drop by up to 83 per cent in just two decades. These nicotine-based insecticides target the nervous system just like nicotine does, so they are toxic to midges and other tiny creatures that live in the lake. As a result, the run-off led to a dramatic collapse of fish stocks.

Neonics are often applied directly to seeds before crops are planted, but the chemicals remain in the plant and contaminate its nectar and pollen. This decimates populations of bees and other pollinating insects, and leads to a decline in the animals higher up the food chain. In the Netherlands, that's farmland birds such as swallows, starlings and tree sparrows. In Canada, neonics contaminate the milkweed plants that surround the corn crops – millions of monarch butterflies that migrate to Mexico every year lay their eggs on the milkweed, and neonics dramatically reduce their chance of successfully hatching. The EU ban of neonics on farmland came into force in 2018, but doesn't apply to use in rivers and the sea, so some fish farmers could start using the insecticide imidacloprid to kill sea lice on caged salmon.

Scientists have proved that pesticide levels could be cut without reducing yields on most crops and without losing profit.[27] A UN report that criticized the global corporations that make pesticides stated that 'food that is contaminated by pesticides cannot be considered as adequate food' and concluded that pesticides are not necessary to feed the world.[28] The authors urged for a transition away from 'pesticide-reliant industrial food systems' towards regenerative farming that does not damage the environment. And some countries are more ahead of the game than others. On the eastern edge of the Himalayas, the country of Bhutan claims to be the world's first 100 per cent organic country and the Indian state of Sikkim has followed suit. In Switzerland, where Syngenta, one of the world's largest producers of pesticides, has its headquarters, a referendum was held in summer 2021 to vote for or against the use of synthetic pesticides in farming and in public green spaces. While a ban was rejected, the result of this vote wasn't a foregone conclusion. The referendum was instigated by a citizen-run campaign and just goes to show that radical change is perhaps possible.

There has been a revolution against all pesticides in Cuba, but less of an intentional one. Because it became so difficult to get hold of agricultural chemicals in 1991, when the Soviet Union collapsed, 10,000 Cuban farmers changed their ways and formed organic cooperatives. This was driven by an urgent need to produce food, not a desire to reduce impacts on the environment, but had huge positive knock-on effects. Cuba is now pretty self-sufficient when it comes to fruit and vegetables, with thriving networks of urban farms that raise insects for biological pest control and use worms to break down organic waste and create vermicompost. Around the coast, there's no agricultural run-off, and the mangroves, wetlands, seagrass beds and coral reefs are healthier as a result.

Breathe easy

When an inquest found that excessive levels of nitrogen dioxide and PM2.5 exacerbated the acute asthma that led to the death of nine-year-old Ella Adoo-Kissi-Debrah, who lived close to the busy South Circular Road in South London in the UK, she became the first person to have air pollution officially cited on her death certificate. Air pollution is the world's largest single environmental health risk, resulting in the deaths of an estimated 7 million people annually, yet it remains an invisible killer and there is no safe threshold for air pollution.

Outdoor air pollution is well documented in the US and Europe, but less so across the developing world. Countries with better monitoring will show higher rates of air pollution but that doesn't necessarily mean that other countries are problem-free. Air pollution isn't just associated with busy cities. VOCs released during the use of pesticides and other agricultural chemicals are a serious problem in rural areas too. Air pollution could have major implications over oceans, especially near shipping lanes, but this hasn't been studied as much as on land.

Human activities since the industrial revolution have resulted in the excessive emission of many air pollutants into the atmosphere, from coal-burning power plants, mines and factories to waste incineration, transport and agricultural practices. One of the POPs listed by the Stockholm Convention on its initial dirty dozen list, dioxins are one of most toxic unintentional by-products of these industrial processes. These endocrine disruptors and carcinogens are created during smelting, pesticide production and manufacturing processes, such as chlorine bleaching of wood pulp and cotton, and to some extent during cooking. One of the biggest contributors of dioxins is waste incineration, which results in incomplete burning. Once released into the air, dioxins settle in the soil and water, and accumulate in the food chain. More than 90 per cent of human exposure is through the food that we eat.[29]

In addition to dioxins, the air that we breathe contains a mix of pollutants such as carbon monoxide, sulphur oxides that pour back down as acid rain, nitrogen oxides, plus invisible solid particulates. This particulate matter or PM is defined by its size. PM10 (10 micrometres or smaller) and PM2.5 (2.5 micrometres or smaller) are found in soot, dust and tyre rubber, plus brake wear from vehicles, traffic fumes and smoke from log burners, which all mix with liquid droplets to form an aerosol. Measurements of PM2.5 in the air show that fifteen of the twenty most polluted cities are in India and the smaller the particles are, the greater the health risk they pose, just like smoking. PM2.5 can enter the bloodstream, reach other organs, such as the brain or heart, and cross the placenta of pregnant women. Although PM2.5 tends to be concentrated in urban and industrial areas where it can form a haze, PM2.5 also pollutes our homes.

THE WORLD'S TALLEST AIR PURIFIER?

Inspired by the dense smog he experienced while living in Beijing, Dutch artist Daan Roosegaarde collaborated with Bob Ursem, a nanoparticles expert at the Delft University of Technology, Netherlands, to design a smog-free tower that cleans the air. Powered by solar energy, the 7-metre-tall tower acts like a vacuum cleaner, filtering 30,000 m^3 of air every hour and removing PM2.5 and PM10 particulates. Smog-free towers have been installed in South Korea, China, Poland and Rotterdam. Every month they open up like a spaceship so that pollutants can be removed. Carbon in the smog particles is transformed under high pressure into diamonds. Roosegaarde revealed that the air around the tower is 55 to 75 per cent cleaner than the rest of the city. 'It's a great local solution, but it's not about creating a whole clean-air city – that needs green energy and electric transport,' explained Roosegaarde.

In polluted areas that aren't well ventilated – underground car parks in the city centre, for example – carbon-monoxide poisoning can kill. Once inhaled into the lungs and absorbed into the bloodstream, carbon monoxide prevents the blood from carrying oxygen, resulting in breathlessness, dizziness and unconsciousness. It's notoriously difficult to detect because it's colourless and doesn't smell, and can happen inside homes when fuels such as wood, gas, oil or coal don't burn fully. So it's absolutely crucial to ensure proper ventilation, but it's always worth considering the air quality outside too. If you live by a motorway or close to factories, it might make sense to filter the air indoors instead of opening the windows. The importance of indoor air pollution is massively underestimated. Levels of air pollutants can be two to five times higher indoors than outside.[30] Around 3 billion people, mostly in low-to-middle-income countries, still cook and heat their homes with fuels such as charcoal, wood and coal, which are often burned in very inefficient stoves indoors. This results in spikes of PM2.5 and the release of VOCs and other pollutants that can cause conditions such as pneumonia, stroke, heart disease and lung cancer.

Nitrogen oxides (NOx – pronounced 'knocks') are one of the most notable indicators of air pollution. Released from traffic fumes, power plants and wildfires as well as during the combustion of fossil fuels in buildings as a source of heat, NOx can add to the unnatural extra influx of nutrients in water that reduce oxygen levels in rivers, lakes and oceans, resulting in dead zones. But many cities are finding ways to keep these fumes at bay in busy hot spots. Europe already has more than 250 clean-air zones which can drastically reduce NOx emissions when implemented alongside incentives for using electric zero-emission vehicles and efficient, affordable public-transport systems. Filtered traffic systems have been put in place in Dutch cities such as Amsterdam and Utrecht, while low traffic 'superblocks' that restrict access to cars are being pioneered in Barcelona, Spain. Research shows that if Barcelona rolled out all of its 503 planned 'superblocks', nitrogen dioxide emissions would fall by 24 per cent, 667 premature deaths could be prevented and 65,000 people would shift to public transport and active travel such as walking and cycling.[31] With larger designated pedestrian areas, greater investment in cycling infrastructure and car-sharing initiatives, air pollution in cities can be dramatically reduced.

When the Chinese megacity Shenzhen introduced more than 160,000 electric buses and 22,000 electric taxis, CO_2 emissions fell by 48 per cent.

While damage to the ozone layer in the upper atmosphere is bad news, the formation of ozone closer to the ground is another problem associated with air pollution and one that is often overlooked. When nitrogen oxides and VOCs emitted from vehicle exhausts react with heat and sunlight, ozone creates harmful ground-level smog that harms human health and damages crops because it impairs a plant's ability to function and photosynthesize. In urban, industrial locations such as New Delhi, India, flights often get cancelled due to reduced visibility. In densely populated urban areas such as New York City, VOCs from personal care products play just as big a role in driving the high ozone pollution levels as VOCs produced by fossil fuels.[32]

Poland has some of the most polluted air in Europe and is still largely dependent on coal-fired power. In Kraków, where the smog can be particularly bad during the winter when the use of coal-burning stoves is at its highest, three university friends invented air-pollution sensors that can be fitted to public buildings to track levels of key pollution markers such as nitrogen dioxide and particulate matter. 'Polluted air is a plague on our health,' said Wiktor Warchałowski, co-founder of the clean-air app Airly. Thousands of air sensors have now been fitted globally since Airly began in 2016, and this information is then fed into an interactive online map so users of the Airly app can see readings of real-time data in more than 600 cities worldwide, from Bucharest to Berlin and beyond. AI-driven algorithms predict air pollution for the next twenty-four hours with 95 per cent accuracy, helping to pinpoint hot spots that local authorities should be tackling and empowering citizens to better plan their routes for outdoor exercise.

Pollution isn't fair

Chemophobia – a fear of chemicals – isn't always rational because our culture is so heavily influenced by marketing. We might panic about which regulated chemical ingredients are listed on the packaging of our cosmetics, but forget about the heavy metals that were used to mine the gold ring we wear or to tan the leather sofa we sit on. Pollution causes more deaths than war or hunger or malaria, but sometimes the most serious impacts are invisible and far away. In the developed world, it's easy for our perspective to become quite skewed, yet we continue to buy things that are creating pollution problems elsewhere in the supply chain. That comes with serious implications. An estimated 22 per cent of the 3.45 million deaths caused by PM2.5 particulates could be attributed to goods and services produced in one region, but which are consumed somewhere else thanks to international trade.[33]

Toxic chemicals pose different risks to people and the environment, either at the manufacturing stage, during use or after disposal. People working in chemical manufacturing, mining or informal recycling might rely on these industries as their only source of income. In places such as Chad, Central African Republic, North Korea, Niger and Madagascar, poor water sanitation and contaminated indoor air contribute to the highest death rates attributed to chemical pollution.[34] Generally, smaller, poorer countries are much worse off than high-income countries, and within any population, children are normally the most affected.

In India, where lead poisoning looms large, many people rely on unregulated or informal recycling of lead-acid car batteries as their primary source of income. But it's a hands-on process that releases toxic chemicals directly into the environment, as liquid is poured out onto the land and fumes are emitted during smelting. Lead-contaminated waste ends up in soil and crops, water systems and in the air, while lead blood levels in the recyclers on the front line are particularly high as a result. Lead contamination also spreads around the world in exported fruit, veg and spices.

The international charity Pure Earth specializes in toxic pollution clean-ups in low-to-middle-income countries, with a strong focus on lead as the most serious, and easily preventable, health concern. In India, while remediation of lead contamination is a key step, a systems shift is underway too. As it stands, informal recyclers don't have to pay tax so they can massively undercut formal recyclers when dead car batteries are put up for auction. Now, Pure Earth is working with the Indian government and the Material Recycling Association of India to exempt recyclable materials from this tax and create a more even playing field so that all batteries are sent to formal recycling operations and lead poisoning can be stopped at source. Once the system has changed, informal recyclers will be more motivated to seek out alternative, safer options of making a living.

Clean-ups only really happen when appropriate laws are enforced, but remediation alone is never the answer. The phasing out of chemicals of concern and the design of safer alternatives is the way forward. Pollution stops when we turn off the tap but as the climate crisis accelerates, and floods, hurricanes and wildfires become more frequent, there's an increasing risk that toxic chemicals could overflow from designated decontamination sites to surrounding inhabited communities. Right now, it's time to redress the balance, consider whether our use of certain products is essential and reduce our chemical consumption accordingly, while looking at the pollution landscape through a global lens.

PART TWO

How to Go Toxic Free

4

In the Bathroom

Bathrooms are a clean haven, aren't they? We wash away the grime and emerge feeling fresh and renewed. But have you ever considered exactly what's in the products you use and where they might go afterwards?

From the shower gels, shampoos, conditioners and lotions lined up on the windowsill to the toilet cleaners or mildew spray lurking under the washbasin, there is such an enormous range of products in our bathrooms. Not to mention the medicines, make-up and hygiene items, many of which get thrown away once expiry dates have been and gone. If you empty your bathroom cabinets and take a look at the pile of pots, bottles and tubes, you might be surprised by the sheer volume. How much do you use on a daily basis? Decluttering could not only help streamline your shelves and save you money in the long term, but also reduce your exposure to chemicals. If you find unopened gifts that you'll never use, donate them to a charity shop. If you discover lots of out-of-date items, avoid pouring them down the washbasin or flushing them down the loo. Check the label because some dangerous chemicals such as disinfectants need to be taken to the local household hazardous waste disposal site rather than being thrown away in the bin. Think particularly carefully about tablets and other pharmaceutical drugs that are no longer required. To ensure their safe disposal, it's always best to return them to your local pharmacist.

Of course, there's no sense in just binning everything. Use up what you do have and when you next go shopping, take time to read the labels, make more eco-conscious choices and buy fewer products. Reducing the number of bottles in your bathroom not only creates more space, but is a great way to limit wastage – and that alone is a huge, positive step.

Many years ago, I had a habit of mindlessly washing away my used daily contact lenses down the plughole each night. It never occurred to me back then that doing this might have an environmental impact. Thankfully, I now know differently. Instead, I save them, along with all the packaging, and return them to the opticians for recycling via a global recycling platform called TerraCycle, which specializes in turning previously non-recyclable post-consumer waste into new products.

It's time to scrub up on the science, consider long-term exposure to ingredients and make a conscious choice about which products to use every day on our bodies.

It's all too easy to forget that the water washed down each plughole or toilet in our bathroom eventually connects to the ocean. Contact lenses and other small unflushable items like cotton-bud sticks, tampon applicators and dental floss can easily sneak through the filtering screens of water-treatment plants, and end up in the sewage system and the waterways beyond.

From cleansing wipes to toilet wipes and baby wipes, language on labelling can be incredibly misleading. Claims that products are 'biodegradable', 'natural' or 'flushable' aren't regulated and most won't ever degrade in sewer conditions but instead cause contamination, blockages and hazardous fatbergs. In the UK, a new 'Fine to Flush' symbol introduced in 2019 indicates that the product has been stringently tested and certified as compostable by Water UK, a membership body for water and sewerage companies in Britain. Some retailers are getting their wipes rigorously tested and certified, but many aren't so governments need to ban plastic wipes altogether. If products don't show the Fine to Flush logo or a certified equivalent, bin them. Better still, opt for reusable, washable wipes. We need to take a long-term view that goes beyond our fleeting consumption of a single-use product.

With the more invisible problem of microbeads, the tide is turning. These tiny, spherical, plastic particles, designed to cleanse, unclog pores, help with exfoliation or remove plaque from teeth, were purposefully added to toothpastes, face scrubs and shower gels. Once used, they're rinsed off our skin and washed down the drain, but are too small to be retained and filtered out by the wastewater treatment plants. Thanks to manufacturing and sales bans of plastic microbeads used in cosmetics and personal health-care products in the US, Canada and the UK, plus the voluntary removal of microbeads from rinse-off products across Europe's cosmetics industry, the vast quantities of microbeads washing into the waterways is being reduced in some parts of the world.

China has plans to ban microbeads in cosmetics and stop the sale of existing stock by the end of 2022, and many other countries are now considering taking similar action. The UN has recommended a worldwide boycott and, as it stands, the European Commission has pledged to ban microplastic from cosmetics and many other products. But new laws aren't yet in place, and individual responsibility is really key to making simple changes that can make a huge difference to our environmental impact.

Some natural alternatives to these microbeads do already exist. Beads can be made from oil from the jojoba plant, as well as from salt, coffee and oats. Scientists at the University of Bath in the UK are developing a biodegradable, renewable alternative from cellulose that could be sourced from the waste of the paper-making industry.

Glittery make-up is effectively sparkly microplastic and even so-called biodegradable glitter won't decompose in cold seas. Instead of washing it off into the sink, try to wipe off any make-up with a pad, then throw that pad in the bin. Investing in reusable organic cotton or bamboo pads will prevent waste, but these have to then go in the washing machine where any residues get washed away – so, come full circle and it's plain to see that we need to look at the bigger picture. What is the real cost of buying and using products that have a knock-on effect on our environment and potentially our own health?

Beauty and the beast?

Cosmetics is big business. By 2025, this sector is expected to be worth $758 billion globally.[35] Unsurprisingly, there's a lot of scaremongering as cosmetics brands attempt to outdo each other, but be assured that chemicals have been hazard-assessed for personal use whether ingredients are natural, organic or synthetic. But this often won't factor in the cocktail effects of multiple substances or long-term exposure, so a certain level of common sense is required. More than 16,000 ingredients are listed by the International Nomenclature of Cosmetic Ingredients (INCI), which standardizes labelling names for chemicals in the EU, US, China, Japan and many other countries, and labels nanos as such too. Regulations do vary, though, with EU laws restricting the use of chemicals of concern far more than elsewhere. In the EU, more than 1,300 chemicals have been banned or restricted for use in cosmetics.[36] In the US, that number is just 11.[37]

Many regulations lag behind those in the EU. Formaldehyde – a probable carcinogen – is restricted in the EU and banned in Japan, but in the US it gets added to keratin hair straighteners, and formaldehyde-releasing preservatives are included in products like shampoo, eye shadow and nail varnish.

Some progress is being made, though. In September 2020, California became the first US state to ban twenty-four toxic chemicals from personal care products and cosmetics. This Toxic-Free Cosmetics Act included the ban of formaldehyde, PFASs and mercury. Manufacturers won't want to produce different versions of their products for different states, so this could have a nationwide impact as they will meet the strictest standard.

TOP TEN SHOPPING APPS

A quick scan of the barcode is an easy way to check ingredients in cosmetics, beauty and personal care products via these apps:

1. Detox Me (Silent Spring Institute, US)
2. Chemical Maze (Australia)
3. Skin Deep (Environmental Working Group, US)
4. Giki (UK)
5. Picky (South Korea and global)
6. Beauty Evolution (China)
7. Think Dirty (Canada and global)
8. Yuka (Europe)
9. Clearya (Israel and global)
10. INCI Beauty (France and global)

As well as formaldehyde, nail varnish can be made with other highly volatile chemicals such as toluene, a solvent that can result in dizziness, headaches and skin rashes. Shellac and acrylic nails are harsh on your nails, so you'd be better off filing, buffing and showing off well-manicured natural nails, then using nail varnish for special occasions. Most nail-varnish remover contains acetone, a highly flammable liquid that can cause eye, skin and throat irritation. Non-acetone removers usually contain ethyl acetate which isn't ideal either, so use it sparingly.

Almost a third of cosmetics, from hair styling products and nail polishes to face masks and shower gels, contain hidden liquid polymers or dissolved plastics such as carbomer and acrylates, that are not readily biodegradable.[38] While regulations to eliminate microbeads have gained traction, liquid polymers aren't yet clearly labelled or governed, but the mainstream use of these unflushables needs to be addressed.

Heavy-metal make-up is a real issue in countries where regulations aren't as strict as the EU's. Chromium and cadmium are used in the US as colourants in lip gloss and eye shadow. Again, these are classified as carcinogens. Make-up can include titanium dioxide nanos, some red lipsticks contain traces of lead-based dyes, and coal-tar dyes can still be found in US-produced eye shadows even though they were banned in Canada and the EU years ago. PFASs are widespread in make-up, particularly in waterproof mascara, foundations and liquid lipstick, which could easily be accidentally ingested. PFASs are used to make cosmetics more durable, easier to spread and water-resistant, but they aren't in all cosmetics so they can't be absolutely essential.[39]

The packaging on many skincare creams makes sweeping anti-ageing or wrinkle-free claims that are often accompanied by long, complicated lists of chemicals. If the primary ingredient is aqua or water, extra chemicals must then be added as preservatives to prevent water-based products from going mouldy. The impacts of some of these preservatives on skin health is beginning to emerge. Think of this as a comparison between fast food and a meal cooked from scratch, and remember that our skin is an organ – we are what we eat, after all. Just as our diet or the use of antibiotics can imbalance our gut flora, perhaps so too the stuff we use on our skin could imbalance the flora or natural 'good' bacteria that live in our skin.[40] So streamlining your beauty routine can drastically reduce the number of different chemicals you apply to your body. That's less of a problem with oil-based products, such as facial oils or simply extra virgin olive oil, which tend to be less processed with fewer ingredients.

Simplify your beauty routine. Think of your make-up bag like a capsule wardrobe – choose carefully and pick key essentials that you'll use regularly.

When it comes to haircare, the health impacts of chemical ingredients are likely to be more significant for stylists who are exposed on a daily basis. Low-VOC hairspray products are preferable and if you're colouring your hair, be aware that semi-permanent and permanent (or oxidative) hair dyes cause much harsher chemical changes in your hair than temporary dyes that wash out, and all of them get rinsed down the drain eventually. Some are made with coal tars, high levels of ammonia compounds or hydrogen peroxide to strip back the colour, and fragrances. Exposure occurs through direct skin contact and by inhaling fumes during the application process, and darker dyes tend to contain more chemicals associated with possible health risks, such as the carcinogen trichloroethylene (this glue chemical is also found in hair extensions) or the neurotoxicant toluene which has been linked to birth defects and allergic reactions. Studies into

LOOKING FOR VEGAN BATHROOM PRODUCTS?

Check for the internationally recognized vegan trademark if you want products that avoid animal ingredients and don't involve animal testing.

Shampoos and conditioners often contain keratin. Some haircare products contain milk, beeswax (often listed as cera alba), propolis or honey. Glycerine can be derived from animals or plants – it's hard to decipher which from the label.

Animal musk, the secretions from beavers, deer and otters, is used as a fixative in perfumes to make scents last longer. Many large brands now offer vegan alternatives.

Nail varnish sometimes includes shellac, which is derived from the female lac bug, or opalescent pigments taken from mussels, pearls and oysters.

Lipsticks are often made with beeswax and red pigment known as carmine or cochineal, which originates from insects.

Foundations may contain squalene, an oil found in shark livers that prevents moisture loss. Vegan-certified squalene will be sourced from olive oil, rice and sugar plants.

Also look out for gelatine (used for moisturizing properties), collagen derived from animal fats (for anti-ageing), vitamins like retinol (it might include animal-derived palmitic acid) and vitamin D3 (which can be derived from lanolin in sheep wool).

the carcinogenic effects of hair dyes have shown very mixed results, and risks seem to be significantly higher for people exposed to hair dyes on a regular basis through working in salons than for those who just have their dyed hair often.

So if you work in a salon, make sure the room is well ventilated, apply small amounts of a product, wear a mask if using a lot of spray-based items, wash hands before and after use, and wear protective gloves during treatments. Look out for alternative options such as Hairprint, which uses a vegetable extract from velvet beans instead of synthetic dyes. This Californian company was set up by a chemist who developed a new way to colour grey hair that mimics the natural process of restoring the pigment in hair follicles.

Weigh up the risks when treating your family with nit shampoo. Conventional treatments for headlice are pretty potent – they're an insecticide. Regular use of a nit comb may prevent or minimize the need for intensive nit treatment and alternatives made with tea tree oil and coconut oil are available.

Making sense of fragrances

The International Fragrance Association lists approximately 3,100 fragrance ingredients that are used for odour or malodour coverage. A diverse array of fragrance mixtures are made using hundreds of chemicals, many of which emit VOCs and include endocrine disruptors and carcinogens. The terms 'perfume', 'aroma' and 'fragrance' can represent synthetic scents or those based on natural ingredients.

If a product is certified organic by a recognized certifier such as the Soil Association or COSMOS, added fragrances consist of essential oils. But plant-based essential oils can be irritants depending on a person's sensitivity and the dose or concentration. They release VOCs but are generally exempt from being listed as ingredients.

So, who knows what's in a fragrance? There's currently huge debate about the disclosure of fragrance ingredients and the European Commission is discussing labelling additional fragrance allergens on cosmetics products in the EU. Regulation often depends on how a product is used. If it is purely for cosmetic purposes and used to make that person more attractive rather than having any functional or medicinal benefit, they can be regulated differently. In the US, the Food and Drug Administration (FDA) classifies perfumes, colognes and aftershaves as cosmetics, but fragrances used in shampoos, shaving gels and moisturizers aren't governed in the same way.

Many companies keep the contents of their scents a closely guarded secret due to 'commercial sensitivity', but there's a strong argument for mandatory fragrance disclosure – at least then we can find out exactly what it contains. Even if there might not be space on the packaging label for the full list, it could be published online. For simplicity, opt for fragrance-free alternatives wherever possible.

The paraben puzzle

Many bathroom products are now 'paraben-free', but that doesn't mean much until parabens are put into context. First used in foods in the 1920s and now added to cosmetics and pharmaceutical drugs, these synthetic chemicals are widely used as preservatives. They kill pathogens such as bacteria that might otherwise spoil the product and potentially result in conditions such as eye infections in humans.

Common parabens include methylparaben or propylparaben which are used in all sorts of leave-on and rinse-off products, from sunscreens to mouthwash. Derived from benzoic acid, a naturally occurring compound found in fruit such as apples and berries, parabens have a chemical structure that is similar to oestrogen so they can mimic this sex hormone and are classed as endocrine-disrupting chemicals.

Over the past two decades, parabens have received disproportionately bad press compared to other synthetics. In the UK, they are deemed safe and are strictly regulated by the EU. Proponents, including beauty entrepreneur Liz Earle, argue that parabens have been unnecessarily vilified due to their potential association with increased risk of breast cancer. This stems from 2004, when a scientist at the University of Reading, UK, tested a small sample of twenty human breast tumours that showed traces of parabens in each one. Mainstream media reported that parabens used in underarm antiperspirants might be causing breast cancer, but no definite association has actually been made. In fact, the study did not compare healthy breast tissue for traces of parabens; nor could the scientists guarantee that the equipment used in the experiment had been cleaned with paraben-free materials, so contamination could not be ruled out.[41] Interestingly, since this study, parabens have been found in the breast tissue of women who didn't use underarm cosmetics, suggesting that parabens must enter the breast from other sources.[42]

While critics argue that paraben-free products are preferable, proven safe alternatives are limited and the levels used are extremely low. When I spoke to Dr Robin Dodson about parabens she told me that they aren't top of her list of concerns because they are fairly weak mimics when it comes to their endocrine-disrupting properties. Some parabens are considered to be 'pseudo-persistent' so we always have some in our bodies because they're so widely used, Dodson told me. So, despite being easily broken down and excreted in urine, constant daily exposure through the products we use and the food we eat means that levels are forever being topped up and safety testing won't always account for the long-term impact of that.

Sharima Rasanayagam, director of science at Breast Cancer Prevention Partners (BCPP) explained that few studies show a direct causal link between parabens and breast cancer. As part of the BCPP's Campaign for Safe Cosmetics in the US, she told me that there's more evidence that the longer chain parabens do have endocrine-disrupting properties and have been shown to affect breast cancer cells, so there's a plausible link but not a causal connection. Until we know more, consider how you could limit your exposure, perhaps with leave-on products such as moisturizers, especially if you're pregnant or breastfeeding.

Clean up on soap

Rigorously washing hands and surfaces with plain soap and water is the most effective way of getting rid of germs. That physical removal of microbes through scrubbing, rinsing and drying is much more important than the chemistry of what's actually in the product.[43]

Despite this simple fact, the antimicrobial triclosan is still added to lots of soaps, toothpastes, body washes and cosmetics. Since its first use as a hospital scrub in 1972, it has been incorporated into so many products and its use is now widespread. It's even used as a preservative to stop the growth of bacteria in some clothing, kitchenware, furniture and toys. Studies have found worrying traces of triclosan in 97 per cent of breast milk samples from lactating women and in the urine of nearly 75 per cent of people tested.[44]

Antimicrobials like triclosan encourage bacteria to evolve resistance, and as a result bugs become more virulent and harder to kill. Interestingly, there's no difference in efficacy between soaps containing triclosan and those that don't. If a chemical doesn't add any value in terms of function, then why bother using it at all? In 2016, triclosan was banned from US soap products following a risk assessment by the FDA. In Europe, the use of triclosan in cosmetics products is restricted under REACH (which stands for Registration, Evaluation, Authorization and Restriction of Chemicals), the strictest set of chemical laws in the world. After finding that triclosan is readily absorbed into the human body, scientists have recommended that its use in consumer goods and personal care products should be re-evaluated. Triclosan has been discovered in fish, earthworms and dolphins, and numerous investigations have reported that triclosan acts as an endocrine-disrupting chemical and it has been linked with osteoporosis.[45]

Despite triclosan being declared safe for consumer use around the world by Europe's SCCS and the industry-funded US Cosmetic Ingredient Review, some major manufacturers are reformulating their products without triclosan following public pressure. Unilever, which owns over 400 brands and is the largest producer of soap in the world, has dropped it from all its UK products 'to satisfy consumer preferences'.[46] Clearly, further research is required when so many questions remain unanswered.

Brush up on oral hygiene

Toothpaste is made with a foaming agent known as sodium lauryl sulphate (SLS) that is used to create a lather and help remove dirt from teeth. If you've been using a certain brand all your life with no side-effects, there's no need to change, but SLS can trigger oral inflammation and irritation in people with sensitivities.[47] All the more reason for proper rinsing after brushing and for children to only ever use small amounts of toothpaste to brush with.

Some toothpaste contains nanos. Food-grade titanium dioxide (TiO_2) is used as a pigment to make toothpaste white. It could potentially affect the liver once ingested orally and, once in the environment, could accumulate in crustaceans, fish and algae. In France, titanium dioxide has been banned in food since 2020 and some beauty products do actively market themselves on being titanium-dioxide-free.

Natural toothpastes often market themselves as fluoride-free. But fluoride is a naturally occurring mineral that is known to prevent tooth decay and is deemed safe by medical professionals globally for topical use. This is not to be confused with labels listing fluorine-based ingredients, though – these are PFASs which have been found in dental floss such as Oral-B Glide.[48]

While gargling with mouthwash used to be a standard way to get minty fresh breath, short-term use now tends to be only recommended when gum disease seriously calls for it. Many mouthwash formulations have a high alcohol content, plus chlorine dioxide, which works as a whitening agent, antiseptics such as chlorhexidine, detergents, flavouring ingredients and sometimes formaldehyde. Mouthwash could disrupt the healthy bacteria that play an important role in your oral microbiome and thorough brushing should be sufficient.

The talc debate

The talc in talcum powder is a naturally occurring substance consisting of hydrogen, oxygen, silicon and magnesium, known for its softness and drying ability. This mineral comes from crushed talc rocks that are mined from the soil and is used in foundation, eye shadow and lipstick too. There's big debate about the health risks of talcum powder when used in intimate areas for personal hygiene – some evidence proves inconclusive, while some studies link intimate talc exposure to ovarian cancer due to contamination with asbestos fibres. Asbestos is a mineral that is often found naturally in the earth alongside talc, but it's a known carcinogen that can be harmful when inhaled.

More than 19,000 talc-related lawsuits have been filed against Johnson & Johnson and investigations by Reuters and *The New York Times* documented that Johnson & Johnson knew as long ago as the 1970s that the company's raw talc and some of its manufactured products sometimes tested positive for small amounts of asbestos fibres. Following huge campaigns, Johnson & Johnson announced in 2020 it would stop selling talc-based powders in North America but continues selling them to the international market. Powders made with corn starch are preferable, especially for babies or for use in intimate areas.

Getting sweaty?

Aluminium and its salts can be used in antiperspirants, oral care products and cosmetics such as lipsticks, foundations and eyeshadows. When used in antiperspirants, it's bound into a gel that sits on the skin's surface to plug the sweat pores. The general consensus from the scientific community is that aluminium as an ingredient in cosmetics and personal care products is non-toxic. The molecules are too big to get absorbed into the skin or bloodstream, it is not bioavailable to the body, and even if it were to be absorbed, the amount we ingest in our food and drink would be much greater.

Despite some media hype, a panel of leading oncologists reviewed fifty-nine pieces of research published between 1994 and 2008 relating to antiperspirant and deodorant safety, and concluded that there is no scientific evidence that deodorants or antiperspirants cause cancer.

Sunscreen in the sea

One of the most effective ways to avoid exposure to ultraviolet (UV) radiation – the main cause of skin cancer – is sunscreen. Chemical sunscreens work like a sponge, absorbing the sun's rays, while physical or mineral sunscreens act as a shield, reflecting the sun's rays away from the skin. In physical sunscreens, inorganic UV filters like zinc oxide and titanium dioxide nanos are generally considered safe and unlikely to be absorbed through the skin or into the bloodstream.[49] But when inhaled into the lungs in large doses, titanium dioxide can be a possible carcinogen.

Oxybenzone, globally the most widespread UV filter, is one of the most concerning ingredients because it acts as an endocrine disruptor in rodents and potentially could harm humans too. Strikingly, it has been discovered in human breast milk, amniotic fluid, urine and blood. Oxybenzone contaminates the bodies of more than 96 per cent of Americans according to the Centers for Disease Control and Prevention. The Environmental Working Group (EWG) called for further research after finding this allergenic chemical in 40 per cent of non-mineral sunscreens.

Sunscreen chemicals impact the ocean too, from sex changes in fish to coral bleaching. Between 6,000 and 14,000 metric tonnes of sunscreen is estimated to end up washing into coral reefs around the world every year.[50] One drop of oxybenzone in six and a half Olympic-size swimming pools is enough to bleach coral.[51] Destinations such as Hawaii, the Republic of Palau and the Republic of the Marshall Islands in the Pacific Ocean are banning the use or sale of sunscreens that contain oxybenzone and other chemical UV

TOXIC-FREE SUN SAFETY TIPS

- Choose a physical sunscreen with inorganic UV filters.
- Avoid sunscreens that contain oxybenzone and octinoxate.
- Avoid sunscreens with insect repellent to prevent these added ingredients being absorbed.
- Use sunscreen creams rather than sprays or powders which could be inhaled.
- Avoid sunscreens that contain PFASs.
- Avoid applying sunscreen to lips or sunburnt and broken skin to minimize absorption of nanos.
- Don't be fooled by 'reef-safe' claims – unregulated claims are rife.
- Use sunscreen in combination with clothing, hats, sunglasses and shade.

filters. Once safer alternatives that won't compound the problem become more widely available, this risk could be an easy one to eliminate.

Bottoms up

Globally, we use a lot of toilet rolls. People in the US lead the way with an impressive 141 toilet rolls used per person on average every year. That figure is 134 rolls in Germany, 127 in the UK, eighty-eight in Australia and just forty-nine in China because across South East Asia, washing with a bidet shower is a more common practice than using paper.[52] To make sure the planet doesn't get a bum deal, buy recycled toilet paper made with less bleach than that made from virgin wood pulp.

Babies can wear up to an eye-watering 4,000 single-use nappies in the first three years of their life, most of which are made from petroleum-based plastic that takes centuries to fragment. As well as fragrances that can trigger allergies and endocrine-disrupting phthalates, low levels of the weedkiller glyphosate have been found in nappies. In 2019, the French Agency for Food, Environmental and Occupational Health & Safety (ANSES) reported that chemicals in babies' nappies exceeded safety levels and could pose a threat to infant health.[53]

To reduce chemical exposure, choose fragrance-free nappies. Eco-friendly brands of FSC-certified disposable nappies include Bambo Nature (produced in Denmark, with Asthma Allergy Nordic and Nordic Swan Ecolabel certifications), Naty by Nature Babycare (made in Sweden, these are chlorine-free, perfume-free and have been certified by the Swedish Asthma and Allergy Association) and Seventh

Generation (the first US Department of Agriculture Certified Biobased nappy). The overall impact of using reusable cloth nappies, preferably made with certified organic cotton, can be lower than disposables if you don't tumble dry them. In many European countries, Singapore and Australia, nappy libraries run helpful demonstrations and loan kits of cloth nappies, so you can try before you buy.

Toxic-free periods

A woman menstruates for approximately thirty-eight years, needing on average 11,000 disposable tampons or pads during her lifetime.[54] But the daily disposal of millions of single-use menstrual products and their packaging is a huge environmental concern. While most goes to landfill, a substantial amount gets thrown down the toilet, perhaps because of the stigma associated with periods. These items aren't always filtered out by water treatment plants, and sanitary items are the fifth most common item found on Europe's beaches.[55]

Like nappies, the plastic component of this sanitary waste never biodegrades. Up to 90 per cent of menstrual pads and 6 per cent of tampons have been estimated to be plastic.[56] The plastic layers in pads are known as non-woven material and wood pulp makes up the rest of the pads, while tampons include cotton, rayon (made from dissolved wood pulp regenerated as cellulose fibres) or a mix of both. The individual wrapping of items such as tampons gives the impression they are sterile, but unless they're classed as medical grade this extra packaging is totally unnecessary.

More reusable menstrual products are now available – from washable period pants and pads to silicone menstrual cups that can be used for up to ten years. But it's not just a case of reducing plastic pollution. Tampons, pads and menstrual cups are put into or located next to one of the most absorbent and highly sensitive areas of our bodies, alongside an alarming number of toxics, from fragrances to glyphosate residues from conventional cotton-farming practices.

> **Toxic chemicals don't belong in menstrual products. Period.**

Soon after the Women's Environmental Network (Wen) was founded in 1988, it launched a campaign to reduce chlorine bleaching to purify the cotton and wood pulp used in menstrual products and nappies. During processing, brown wood pulp is bleached to make it look white and to kill off any microbes. Wen has been persuading producers to switch to using processes that produce less dioxin, a persistent organic pollutant (POP) made during the chemical processing. Dioxins build up in our fat tissue and have been linked to reproductive disorders, damage to the immune system and cancer. Some 'chlorine-free' claims are misleading because only certified organic products made with 'totally chlorine-free' process won't release any harmful dioxins. So, despite some changes in bleaching practices at wood-pulping mills, both dioxin and chlorine are still found in many tampons and pads.

Alexandra Scranton, director of science and research at the US-based Women's Voices for the Earth (WVE), worries that there's still so much to learn about how menstrual products affect women's sensitive vaginal tissue and their reproductive health. 'Manufacturers just don't have a sense of what these chemicals are really doing to your uterus or vaginal bacteria or even whether fibres are shed from tampons during use. We really need to know how this all works to minimize that risk for everyone,' Scranton told me.

TOXIC-FREE PERIOD TIPS

- Switch your conventional tampons, pads or panty liners to 100 per cent certified organic cotton disposables.
- Avoid products with synthetic fragrances, added lubricants and odour-neutralizing technology.
- Change one thing at a time – perhaps try out washable period pants or reusable pads at night.
- Put disposable pads, tampons and applicators in the black bin, not the toilet.
- Consider using a menstrual cup – made of medical-grade silicone, it costs just 16p per cycle over the course of a decade.
- Demand to know more – contact manufacturers to ask them to declare the full ingredients list for the products you buy and use.
- Talk to your friends, nieces, daughters and sisters about toxic-free alternatives and empower others to make healthier choices.

WVE found that both unscented and scented Always sanitary pads emit VOCs that are classified as carcinogens (including chloroform and styrene), reproductive and developmental toxicants.[57] In South Korea, a country where the toxicity of menstrual products is often not discussed, scientists used the same protocol as WVE to investigate VOCs in menstrual pads and warned of the need for precautionary measures. As a result, some ingredients are now disclosed on menstrual product labels in South Korea.

In some period underwear and menstrual pads made with anion strips or other ion technology, nanosilver is added as an antimicrobial. But is nano a silver bullet? Not only can these nanos spread into the environment once washed, they're being used on a sensitive and highly absorbent part of the body. The vaginal area naturally has a healthy population of beneficial bacteria, so antibacterial menstrual products really aren't ideal. In the US, the FDA Office of Women's Health is funding research into the toxic potential of nanosilver in feminine hygiene products. Until that research is complete, manufacturers are posing an unknown, and unnecessary, risk to women.

In addition to personal care products, bathrooms attract an astonishing array of washbasin unblockers, mildew sprays and toilet cleaners. The picture of a dead fish on bleach bottles explicitly indicates that washing this liquid down the drain harms aquatic life. We need to wash away the dirt, but a biocide that kills all bacteria, good and bad, isn't essential at home. Plus, the heat and humidity of the bathroom can increase the chance of chemicals seeping out into the air so think how and when you use products in the bathroom.

TOXIC-FREE TAKEAWAYS

- Most personal care products end up inside your body, in the bin or down the drain and get left to the water authority to deal with, so choose carefully.
- Only ever flush the three Ps down the loo – pee, poo and paper – and be wary of wipes.
- Context is key – exposure to chemicals in tampons that go inside our bodies dramatically differs to that of rinse-off products.
- If ingredients such as fragrance chemicals or nanos aren't listed, ask the manufacturer for more information.
- If you work in a beauty salon, barbers or hairdressers, look for ways to avoid excessive exposure to toxic chemicals on a daily basis.

5

In the Kitchen

The kitchen is widely pitched as the heart of the home. A place where delicious meals get served up, comforting cups of tea are brewed, and stories can be shared around the dinner table. But from how you heat your food to the products you use to clean the work surfaces, toxic chemicals can abound in this room. The good news is that there are lots of simple ways to make your kitchen safer, healthier and less polluting all round.

Cooking up a storm

Aside from the dangers posed by sharp knives and hot pans, cooking can be dangerous. Whenever you cook, aerosols that contain hundreds of chemical compounds are generated. Women and children are most at risk of breathing these in and according to the WHO, 3.8 million people die prematurely every year from household air pollution from cooking. Cooking is a major source of air pollution in densely populated megacities of China and India.[58]

Cooking appliances vary enormously, as do cooking methods, so it's hard to generalize. Cooking with gas stoves, grills and ovens can be a source of carbon monoxide, nitrogen oxides and particulates which are linked to health issues including heart problems, cancer and diabetes. The highest levels of PM2.5 containing fine soot and tiny particles of animal fat or cooking oils come from roasting food in ovens and cooking food on gas hobs, while slow cookers release fewer particulates. While boiling vegetables and roasting turkey for a Thanksgiving meal at a test house in Austin at the University of Texas, PM2.5 levels rose to 200 micrograms per cubic metre for one hour – more than the 143 micrograms per cubic metre found in Delhi, the sixth most polluted city.[59]

For vegetables, steaming or microwaving uses much less energy than boiling on the hob. Microwaves have been shown to destroy the nutritional content of food, but so too does any oven cooking to some extent, while boiling veg results in some nutrients leaching out into the water. Microwaves are energy-efficient because they don't need pre-heating and when switched off, the heating stops straight away, so there's no wasted heat.

The greatest gains to be made are in low-to-middle-income countries, though, where open fires are still widely used for cooking and ventilation is severely lacking. Moving away from combustion fuels to clean stoves run on electricity or solar power is the ideal solution.

A sticky situation

Teflon is used to make kitchen pans, utensils, baking tins and equipment like waffle makers and sandwich toasters non-stick. At high temperatures, harmful gases can be released.

TOXIC-FREE COOKING TIPS

- Try to avoid burning food – frying, roasting, toasting and grilling result in the emission of particulates.
- If you have a cooker hood, choose the back rings of the hob whenever possible to capture maximum pollutants.
- Open kitchen windows and keep a wall-mounted extractor fan on at its highest setting during cooking and for a few minutes afterwards.
- Wash or change the filters regularly on your extractor fan or cooker hood.
- If you need to buy a new hob, opt for electric over gas.
- If you use a gas cooker, install a carbon-monoxide detector in your kitchen.
- Never preheat non-stick cookware to a high temperature.
- If you're buying new cookware and baking equipment, choose stainless-steel, cast-iron, enamel or ceramic-coated options and oven-safe glass that are PFAS-free.
- Instead of greaseproof baking paper, which can sometimes contain PFASs, add butter and flour dusting to baking tins.
- Dispose of any non-stick cookware (especially old or damaged pans) at your local recycling facility.

This has been linked to cases of 'Teflon toxicosis' whereby fumes result in the poisoning and subsequent death of pet birds. The Teflon website advises keeping your feathered friends out of the kitchen while cooking and to avoid preheating cookware on high heat settings.

The manufacture of Teflon is toxic. Until 2015, Teflon, the brand name for a chemical called PTFE (polytetrafluoroethylene), was made using a highly toxic and persistent PFAS called PFOA (perfluorooctanoic acid) or C8. The chemical manufacturer DuPont was forced to phase out C8 when a huge scandal revealed that the company had been dumping C8 into the waterways surrounding the factory in West Virginia, US, since the 1950s. Thousands of cases were filed against DuPont after C8 in drinking water was linked to six diseases including thyroid disease, kidney cancer and testicular cancer.

The attorney Rob Bilott, who has battled against DuPont for twenty years, told me that he wishes litigation wasn't the only option. 'We shouldn't have to go into courts to get clean water and get things cleared up. There are hundreds of cases pending now, but we need to change the laws. We've been waiting for a federal limit on PFOA in drinking water since 2001. The manufacturers should take the burden, not the people being exposed. The whole system is flawed.'

So what's changed? While production and use of C8 has been phased out across Europe, Japan, US and Canada, the problem has shifted elsewhere to countries with fewer restrictions such as China, now the largest producer of PTFE.

GenX, the chemical now used to make most Teflon, isn't necessarily any safer. Smaller, replacement C6 and C4 compounds were hoped to be less toxic and less persistent but they are just as stable, and could bioaccumulate more easily. Because of these regrettable substitutions, opting for

PFOA-free products isn't necessarily a safe bet. Because it would be so impossibly slow to evaluate thousands of PFAS chemicals individually, PFAS should be limited to essential use only, when no safer alternative is available and when the benefit truly outweighs the risk.

Look for kitchenware that's PFAS-free. Belgian brand GreenPan makes cookware using a ceramic non-stick coating called Thermolon, which is derived from silicon dioxide – that's sand. A sprayable solution of very fine particles is produced, sprayed onto roughened surfaces of pans and hardened in the oven without adhesives.

Tap, tap

PFASs in drinking water still isn't regulated enough yet. In China, more than 100 million people in sixteen cities are being supplied drinking water with PFAS levels that exceed a safe capacity of an estimated one part per million.[60] More than 200 million Americans could have PFASs in their drinking water, with some of the highest levels in major metropolitan areas such as Miami, Philadelphia and New Orleans. But there's no national standard for PFASs in drinking water in China or the US. The new PFAS Action Act of 2021 calls for the EPA to better regulate PFASs in drinking water, determine whether all PFASs should be classified as hazardous and to introduce labelling that signifies whether products are PFAS-free or not. A new Water Affordability, Transparency, Equity and Reliability (WATER) Act will provide a $35 billion annual trust fund for states to repair facilities, filter toxic chemicals like PFASs from drinking water and replace old lead service lines that can result in poisoning of the water supply.

Traces of agricultural chemicals can end up in tap water. The herbicide atrazine, used widely on US corn crops, has been found in 94 per cent of water tested and has been linked to cancer, birth defects and infertility, plus it can turn male frogs into females that produce viable eggs. Meanwhile, nitrates end up in the water system from agricultural run-off that contains nitrogen fertilizer and manure. This contaminant is especially harmful to babies and pregnant women. Even at levels far lower than legal standards, nitrates can result in low birth weight and an increased cancer risk. High levels could cause blue baby syndrome, a potentially fatal condition whereby nitrates stop haemoglobin in the blood carrying enough oxygen.

CFCS: A SUCCESS STORY?

In 1987, the Montreal Protocol, arguably the most successful global environmental legislation of all time, came into force, banning CFCs (chlorofluorocarbons), chemicals that had been widely used as refrigerants and in aerosol sprays since the 1920s, and regulating almost a hundred other synthetic ozone-depleting substances. But despite production of CFCs reportedly reaching almost zero in 2006, CFC emissions have unexpectedly started to increase again. These emissions come from accidental by-production in chemical manufacture, illegal CFC production in East Asia and old, landfilled fridges and air conditioners or decommissioned buildings, so better access to disposal facilities is essential.

Depending on where you live and which chemicals are most commonly found in drinking water locally, it may be worth using a water filter. Carbon filters reduce contaminants such as lead, or disinfectants including chlorine and fluoride which are used in water treatment plants. At-home 'reverse osmosis' water filtration systems can filter out nitrates, but these tend to be expensive. Instead, it makes much more sense to encourage better legislation that ensures toxic chemicals don't get into drinking water in the first place.

Eat, drink and be toxic free?

Every time we buy food, we are faced with multiple choices aside from just which meal we should cook next. Decisions come down to price, priorities and personal choice. If something is grown locally and not transported as far, it won't have contributed as much to air pollution as food that's been airfreighted. If something is grown in season, it may have been farmed using fewer agro-chemicals, even if not certified organic. Organic, free-range chicken won't contain the same high levels of antibiotics as battery hens. If you're buying a processed ready meal, will you heat it in the plastic packaging it comes in and risk chemicals leaching out into your food?

Food-safety regulations generally focus on pathogens such as E. coli. But if food doesn't have microbiological contaminants in it, does that mean it's safe? Traces of toxic chemicals can cause chronic illness, as highlighted by the Clean Label Project, a US-based award programme that tests products for pesticides, plasticizers and heavy metals that would otherwise go undetected and never get mentioned on food labels.

Even rigorous organic certification doesn't test for heavy metals. If crops are grown in contaminated soil, toxics can end up in our food. In the south-eastern states of America, rice is grown on land with lots of pesticide residues and heavy metals in the soil because it has previously been farmed as tobacco and cotton plantations. Now that rice is sucking up what's in the ground. Jackie Bowen, executive director of the Clean Label Project, tested 564 of America's top-selling baby foods, including pouches, jars, formulas and cereals, and discovered lead in 37 per cent of the samples and cadmium in 57 per cent. There was no association between whether the product was certified organic and its heavy metal concentration, and she found that products containing rice tended to have higher concentrations of lead and cadmium.[61] 'Infant formulas can be the exclusive form of nourishment used during the most vulnerable period of development – these products are highly regulated when it comes to nutritional value,' Bowen told me. By publicly calling out companies on social media, we can demand better labelling, and as more consumers sue brands for being duped or having misleading labels or false claims, Bowen is noticing a shift towards greater transparency.

By showcasing organic and non-organic brands that ensure that their supply chains don't contain pesticides, heavy metals and plasticizers, the Clean Label Project aims to shift food-safety policy to focus more on long-term health. Of course, most products available to eat are compliant in the court of law, but this is just the bare minimum and much more clarity is needed. Ultimately, better environmental policies will enable farmers to grow healthier, safer ingredients and prevent pollutants escaping into the land in the first place.

While organic certification doesn't test for heavy metals, it's still a great measure of a food's toxicity because pesticide use is so widespread. Organic farming isn't free from all agricultural chemicals, but it certainly avoids the most hazardous ones, and pesticides will only ever be used as a last resort. An organic diet can rapidly reduce the amount of glyphosate in the body by 70 per cent after only six days.[62] Even just switching your top three favourite fruit or veg to organic goes a long way to reduce your exposure to pesticide residues that are left on and in non-organic foods. Grapes, oranges, dried fruits and herbs are just some of the foods that, when tested, were found to have multiple pesticide residues in more than 80 per cent of samples.[63]

If organic certifications were ditched, but all non-organic food had to be labelled 'contains pesticides', which would you choose?

In non-organic food, pesticides in Europe are governed by maximum residue levels or MRLs – an upper limit on the concentration of a pesticide residue that's legally permitted on food. But these MRLs don't take into account the mixture of different pesticides used on one crop. The mix of residues varies across food types, and risk assessments should consider the chemical cocktail that an individual might be exposed to when eating one particular food. Washing or peeling non-organic food doesn't necessarily remove chemicals, but the fruit and veg with lower pesticide residues tend to be ones with removable skins or husks such as avocados, onions, kiwi, mangoes and sweetcorn according to Environmental Working Group's Clean Fifteen list for 2021.[64]

Pesticides go beyond fruit and veg too. When it comes to meat and seafood, remember that POPs, including many pesticides, accumulate in fatty tissue, so it's always a good idea to trim off the fat and eat less meat. Pesticides are also used on grains, and farmers are encouraged to spray glyphosate on wheat just before harvesting to help kill and dry the crop, making it easier to store in cool, damp countries. But glyphosate ends up in almost two-thirds of wholemeal bread tested by the UK government. While the levels might not exceed current MRLs, the Soil Association warns that those levels were set before this chemical was classified as 'probably carcinogenic' by WHO and there may be no safe level for human consumption.[65] The only sure-fire way to guarantee that you're eating pesticide-free bread is to buy organic bread or make your own using organic flour.

Even some of the highly hazardous pesticides that have been prohibited for use in the EU can still be produced and exported to other countries. The European Commission aims to implement a ban on these exports by 2023, but for now, loopholes in legislation give way to double standards. Not only are these toxic pesticides causing harm elsewhere, but an investigation by the Pesticide Action Network revealed that some foods being imported back to the EU from places such as China, India, Thailand and Brazil contain residues of banned pesticides and they inadvertently end up on our plates anyway.[66] Exotic fruits such as guavas, goji berries and breadfruit had the highest concentrations of banned pesticide residues. It's easier to avoid these illegal substances if you don't buy as much imported produce, and that's a great way to reduce your carbon footprint too.

Perhaps better labelling would tell us more about the chemicals that are left in and on the food we're buying? More than 1,500 products have been certified glyphosate residue-free by The Detox Project in the US, but campaigns targeted at one specific pesticide are problematic. Firstly, glyphosate can still be used as long as final residues on the end product are not detectable. Secondly, residues of other equally highly hazardous chemical pesticides are not tested for. Thirdly, it gives consumers a false sense of security while providing producers with an unfair marketing advantage for a relatively small cost. Rigorous organic certification is expensive – glyphosate residue-free certification is an attempt to short-cut that.

Pesticide labelling isn't as feasible as it sounds. While it would be useful to know that twenty pesticides have been used to grow the satsuma you're about to eat, there's no way all that information could fit on the label, and the pesticides vary depending on location and time of year too. We can put pressure on retailers to phase out extremely harmful chemicals from their supply chains and support Pesticide Action Network's call for a global treaty on highly hazardous pesticides,.

Some argue that it's impossible to feed a growing population without pesticides, but that's missing the point. Globally, we produce far too much food – as much as a third goes to waste. The real issues are uneven distribution and inefficient practices. One of the most impactful things you can do to reduce your chemical footprint is to cut out food waste in your home. From the irrigation to fertilizers and transport emissions, so much pollution results from growing and rearing the food that we eat, so every bite counts. By consciously choosing to buy, cook and eat foods that have been farmed without the use of highly hazardous pesticides, we can support a system that's better for people and the planet.

Spice it up

The Bengali word 'turmeric' means 'yellow' but the colour of this spice can be a cause for concern. When Stanford University researcher Dr Jenna Forsyth found higher-than-expected lead blood levels in a third of people living in rural Bangladesh, she wondered where this exposure was coming from. It's a country without much industry and leaded gas has been banned there since 1999.

On further investigation, she discovered that lead chromate, a widely available industrial pigment often used to colour plastics and furniture, was being added to turmeric at polishing mills to make it look more appealing.[67] The 'need' for this added colour dates back to the 1980s, when floods during the drying season damaged the turmeric roots. Adding the pigment at the polishing stage helps to remove the drab-looking skin and makes the root inside look brighter, something that has been more of a problem with loose turmeric sold at local markets than packaged spices.

However, thirteen brands of turmeric and curry powder exported by India and Bangladesh, mainly to the US, have been recalled since 2011 because of excessive lead levels. Until Forsyth's work highlighted the issue, public awareness of the link between this pigment and health risks of ingesting lead was pretty non-existent within the communities farming and producing it. Now, Bangladesh's prime minister has vowed to control imports of the pigment and food-safety officers are enforcing the laws that regulate the use of lead chromate with fines to wholesalers selling lead-tainted roots. After 50,000 posters were distributed in 2019 to highlight the risk of using lead chromate and eating it – and encouraging people to buy unpolished roots – lead levels in fresh turmeric samples tested by Forsyth dropped from 50 per cent to less than 5 per cent within just a matter of months. When that exposure stops, blood lead levels should decrease within a few months, according to Forsyth. Lead in spices could be an issue elsewhere in South Asia, and in India – a leading supplier to the global turmeric market – her team found at least one region where lead chromate use is a concern. Now, she's investigating the extent of the problem and looking for ways to solve it as she did in Bangladesh.

Lead in spices is only on the radar of food-safety agencies in a few countries and needs to be much more widely monitored. In our kitchens, the risk is greater with older turmeric products – some people keep spices for years, so the best thing you can do is to buy fresh spices from reputable brands and throw out old ones (in the bin, not the compost).

Pack it in

Aside from what we eat, we need to consider the packaging it comes in too. For decades, a plasticizer chemical known as bisphenol A (BPA) was used as a sealant to line the inside of aluminium cans and metal food packaging. Regulators like the European Food Safety Authority and America's FDA have ruled that the use of BPA in cans does not pose a threat to human health, but many companies including Heinz and Campbell's have started removing BPA from their containers and finding other ways, such as oil and resin substitutes, to maintain food quality and prevent the can from corroding.

Bisphenols can migrate out of plastics into drinking water, food and dust, and it has even been found in sand on beaches. Plastics containing bisphenols are not designed to be heated and higher temperatures speed up the release of chemicals that leach out of the plastic into food. Often, BPA gets replaced with bisphenol S (BPS) and bisphenol F (BPF), which could interfere with hormones in a similar way, so the toxicity of alternatives needs to be investigated. BPA is restricted in baby products in many countries but other bisphenols aren't, so the safest bet for babies is to use silicone teats and heavy-duty glass bottles.

BPA is also used to make the thermal paper that's used for cash till receipts. While touching the odd receipt isn't that big a deal, if you're working behind a till and handling them all day long, regular handwashing is important. More retailers are now emailing receipts instead of printing them. A new app called e.pop (epoppay.com), which stands for 'electronic proof of purchase', collects all your electronic receipts together in one place and encourages consumers and retailers in the US, Canada and the UK to go paperless.

PVC is often used in takeaway containers and it's usually the additives – chemicals like benzene, phthalates and lead stabilizers – that can be released, especially when heated.

Polystyrene or Styrofoam is an environmental nightmare because it gets everywhere and stays there. The liquid styrene that is used to make polystyrene is a probable carcinogen. But switching to alternative materials isn't always any better. One survey found that two-thirds of paper-based takeaway containers were likely to have been treated with PFAS chemicals, which can leach out onto the food.[68] PFASs can be found in greaseproof papers, pizza boxes and bakery bags – anything that's coated to prevent the packaging from getting soggy is likely to contain PFASs, but this won't be labelled. Plus, some of these materials are considered recyclable, but PFASs will end up in that waste stream too.

If you're keen to find out which everyday packaging in your home is likely to contain PFAS chemicals, try this simple test suggested by the environmental charity Fidra: see what happens when you place a droplet of olive oil on an item.

PLASTIC-FREE KITCHEN TIPS

- Avoid heating plastic containers – use glass or ceramic dishes instead.
- If you buy a takeaway, bring your own plastic-free containers along or buy from a restaurant using 100 per cent recycled paper/cardboard that is certified home compostable.
- Don't use plastic drinks bottles that have been left in a hot car because they're more likely to leach out chemicals.
- To reduce packaging, buy fewer processed foods, buy loose fruit and veg, and rinse tinned food in water if possible. Decant things into your own storage jars.
- Switch your cling film for recyclable aluminium foil, reusable beeswax wraps or vegan equivalents.
- Rather than BPA-free products that probably contain other bisphenols, choose reusable options.
- Plastic can be hidden in some products – most teabags are sealed using plastic and at brewing temperatures release billions of microplastics into a single cup. Buy loose-leaf tea or plastic-free teabags.

If it soaks in or spreads out, the chances are the packaging hasn't been made using PFASs. But if the olive oil forms a perfect little round bead, it probably contains PFASs. In 2020, Denmark became the first country to ban PFASs from all food packaging, including baking paper and microwave popcorn bags, and fast-food giants McDonald's and Wendy's plan to remove PFASs in consumer packaging.

Besides packaging, PFASs are used to make stain-resistant kitchen worktops and coatings for sinks. It's impossible to avoid, but manufacturers should be encouraged to find alternatives, especially when it's in direct contact with the food that we eat.

DON'T FORGET THE FOG

When fats, oils and grease (FOG) get washed down the plughole, they can cause blockages after they cool and solidify. Once in sewers, FOG becomes entangled with unflushables like wet wipes, creating monstrous fatbergs that can weigh hundreds of tonnes. It can take several weeks for teams of 'flushers' wearing breathing apparatus and using a combination of high-pressure jets, shovels and pickaxes to dismantle and excavate the nasty congealed matter. New sensors could act as an early-warning system, so that water companies can tackle fatbergs before they become big enough to burst pipes, but that alone won't solve the problem. We all need to be more careful about what goes down the drain. Scrape food scraps into the bin and wipe away residue from cooking pans with a paper towel before washing and avoid putting FOG in the compost. Once cooled, place cooking oil in a sealable container before putting in the bin and use a sink strainer to avoid solid waste washing down the plughole.

Ditch the disinfectant

Kitchens full of toxic cleaning products are as off-putting as grime and our cultural association of pinecone or citrus fragrances with cleanliness is a big misnomer. A hygienic home should not smell of anything at all, but these VOCs are a really noticeable reminder that some household cleaning products can be irritants, resulting in headaches, dizziness and various allergic reactions. Pinene or limonene scents contain terpenes that can react with ozone in ambient air to form dangerous pollutants such as formaldehyde, especially in smoggy places.

Potent solvents called glycol ethers are found in window, surface, oven, floor and multi-purpose cleaners and paints. These endocrine disruptors and possible carcinogens, often listed as 2-Butoxyethanol, 'may damage fertility or the unborn child' according to the European Chemicals Agency. Most oven and drain cleaners contain corrosive substances such as sodium hydroxide (known as lye) that burns through grease and dirt, but makes your eyes sting and can burn the skin or cause vomiting, coughing or pain if accidentally swallowed.

Most disinfectants contain either chlorine bleach (sodium hypochlorite) or quats (quaternary ammonium compounds), and while both of these potent chemicals kill germs for a few hours, they pose serious health hazards. Workers using cleaning sprays and bleaching agents on a daily basis have an increased risk of lung cancer and respiratory diseases, according to data from the European Community Respiratory Health Survey.[69]

Quats can irritate your lungs and skin, and trigger occupational asthma. Quats stay on surfaces after cleaning, so they're not a good idea for use on kitchen worktops where food preparation takes place and when there's a risk that other people and pets can be exposed too. In mice, quats act as endocrine disruptors and affect fertility – more research is needed to find out if this happens in humans. Any cleaning wipes or sprays with ammonium chloride include quats as the active ingredients and should therefore be avoided.

The widespread use of quats in agriculture already contributes to the development of antimicrobial resistance. At the moment, disinfectants still work against household germs, but if used incorrectly that resistance could cause havoc in households too. To ensure that disinfectants remain an effective measure to combat the most dangerous infections in our homes and clinical settings, we must avoid widespread low-level misuse of these chemicals. Many kitchen items such as chopping boards or high-chair trays are sold with extra antibacterial properties that incorporate chemicals such as quats, but reliance on this could result in less rigorous cleaning efforts.

Chlorine bleach is used to disinfect surfaces in medical settings, to treat drinking water and sanitize swimming pools. Aside from toxicity to fish and other aquatic life once in our waterways, it's nasty on so many levels. Firstly, exposure during use – and misuse – is harmful to our health. Bleach releases VOCs, including the gas chloroform, a probable carcinogen, so gloves and a well-fitting mask should be used while disinfecting because it can cause irritation to the mouth, skin and eyes. People with asthma can be more susceptible to flare-ups and children are more vulnerable to the harmful effects of breathing in bleach vapours. If bleach is diluted and used as a spray in a bottle, aerosols or small droplets are produced which are more likely to be inhaled. Secondly, bleach is very reactive, so it's dangerous to mix household bleach with other ammonia-based cleaning agents because this results in the release of really toxic fumes.

Thirdly, the production of chlorine, for PVC and bleach, is incredibly damaging to the environment. It's an energy-intensive process that uses either mercury, asbestos or PFASs to split the salt water or rock salt into chlorine gas and sodium hydroxide. In North and South America, the use of asbestos in chlorine production is still widespread. In Russia and Germany, mercury is used at chlorine plants to convert rock salt from mines into chlorine. In China, the world's largest chlorine producer, PFASs are commonly used. POPs including dioxins and polychlorinated biphenyls (PCBs) are released during these processes.

If quats and bleach are to be avoided, what are the safer alternatives? The best strategy is to keep your home generally clean with normal soap and water, or if disinfectant is really necessary, use isopropyl alcohol. Most supermarket shelves are stacked full of bleach and quats products, so as consumers we need to demand more choice and greater availability of safer, less potent alternatives.

Just like the microbiomes in your gut or on your skin, the microbes on the surfaces inside our homes can have an impact on our health. Regular exposure to certain cleaning agents results in the loss of this microbial diversity, so the remaining ones tend to be more virulent and resistant to antimicrobials. Austrian scientists confirmed that 'human exposure to almost sterile environments should be limited to operating rooms or particular industrial processes in clean rooms', and that in our homes a rich diversity of microbes is beneficial and should be encouraged by regular window ventilation.

Italian microbiologist Elisabetta Caselli has developed the concept of probiotic sanitation which fights germs with 'healthy' strains of bacteria. These counteract the growth of pathogens and decrease the risk of antimicrobial resistance from developing, perhaps by outcompeting other strains.[70]

CONSCIOUS CLEANING TIPS

- Avoid cleaning products with 'caution', 'danger' or 'warning' on the label.
- Use fragrance-free products labelled as low- or no-VOCs, especially for sprays that are more easily inhaled.
- Look for refill options. Check online for in-store refill locations near you, visit your local zero waste shop, buy in bulk or subscribe to a home delivery service that recycles your refill pouches.
- Buy concentrated pods: drop plant-based cleaning capsules or a powdered version into your empty spray bottle, add water and shake to dissolve.
- Choose child-friendly options: preferably a bleach-free and plant-based cleaning spray that gets rid of greasy residues from surfaces, toys, bibs and high chairs.
- Make your own: bicarbonate of soda mixed with sea salt and white vinegar plus a good dose of elbow grease makes a great scrubbing agent. White vinegar with lemon juice produces an effective smear-free, multi-purpose cleaning spray.

Lather up

Until a century ago, most people washed everything with soaps. During the First World War, when the fats and oils required to make washing soaps became sparse, a German chemical company developed synthetic surface-active agents, which became known as surfactants which lower the surface tension between water and grease. Surfactants help remove the dirt from household surfaces, clothes or our skin and today they are the main active ingredients in all household detergents, from dishwasher tablets to laundry powder.

Sodium lauryl sulphate (SLS) is a common synthetic surfactant that helps to produce a lather, emulsify and disperse the dirt; it's widely used because it's cheap, easy to formulate and effective. SLS can sometimes irritate the eyes and make the skin red, itchy, dry or sore. It's not a carcinogen despite having been previously wrongly reported as such by the media, and SLS has been deemed safe for use in household cleaning products. It's usually only used fleetingly and alongside other ingredients not 'neat', but if you're sensitive then avoid direct contact with it.

The world's largest-volume surfactant is called linear alkylbenzene sulfonate (LAS), commonly used in washing powders and washing-up liquids. It's derived from crude oil and only degrades with oxygen, so it won't easily break down in the sewage system. High concentrations of some surfactants can damage the gills of freshwater fish and be toxic to other aquatic creatures. Some surfactants accumulate in sediment and seem to degrade more slowly in marine habitats than in freshwater.[71] Non-ionic surfactants tend to be less toxic to aquatic wildlife and are more commonly used in baby products. But nonylphenol ethoxylates (NPEs) break down into smaller toxic chemicals called nonylphenols

(NPs) which are persistent, so they are being phased out of laundry detergents.

Whenever detergents are packaged into colourful capsules, liquitabs or pods, they become one of the most toxic things found in any home. As well as surfactants, they contain enzymes to break down stains, optical brighteners that make things seem cleaner but won't biodegrade, dyes and fragrances. If a child mistakes one of these bright capsules for a wrapped sweet, this could prove fatal. The first study of the dangers of laundry pods found that in the US in 2012–2013, 17,230 children under the age of six swallowed, inhaled, or suffered eye or skin injuries from these products.[72] They can be extremely alkaline and caustic, burning the skin or the inside of the mouth and airways very quickly, so if you have children or pets, keep them in a locked cupboard.

Residues could be left on crockery, clothes or in the steam that's released when the appliance door is opened at the end of a cycle, so it's worth switching to ones that are approved with the EU Ecolabel or equivalent. Non-bio detergents won't contain enzymes, and laundry balls such as those produced by Ecoegg contain hypoallergenic mineral-based cleaning pellets that release biodegradable surfactants and can be reused.

'Green' biosurfactants called alkyl polyglycosides (APGs) are plant-derived from sugars and made using traditional chemical processing that takes place in laboratories at high temperatures. APGs are biodegradable, safer for sensitive skin and some chemists are finding less energy-intensive ways to produce surfactants by extracting chemicals that already exist in nature. During a fermentation process similar to brewing beer, certain yeasts naturally excrete a mild sophorolipid surfactant so if waste such as cooking oil or reclaimed sugar could be fed to yeasts, production could be much more sustainable.

More brands are using biodegradable surfactants but if it's not clear from the label whether surfactants are petroleum-derived or plant-based, ask the manufacturer for more information. While some ingredients might be fully biodegradable, the final product isn't necessarily so. Many household detergents and personal care products contain synthetic plastic compounds or liquid polymers that are used as thickeners or fillers in liquids, waxes or gels, but won't biodegrade once they get washed away.

Skip the perfumed fabric softeners, many of which are made using rendered (processed) animal fats or 'tallow' from cows, horses and sheep as softening agents to help keep fabrics wrinkle-free. Animal fats, a by-product of the meat industry, are cheap to buy but not everyone wants to wash their clothes using them. Some detergents contain animal-derived lanolin (wool wax produced by glands in sheep skin) or animal glycerol.

Many vegan alternatives are available including fabric conditioner made from rapeseed oil or soluble laundry detergent sheets.

One of the most toxic-free ways to wash your clothes is with organic soapnut shells. These small black berries are harvested from the soapnut tree and then dried. Grown across India, Nepal and southern China, soapnut trees bear fruit for ninety years. In hot water, the shell naturally releases saponin, a natural surfactant that foams in water. A handful of the nuts can be popped into a mesh bag in your washing machine. Alternatively, bring soapnuts to the boil in a pan of water for at least ten minutes and strain the liquid, label it and store in the fridge to use as liquid detergent. Soapnuts are completely biodegradable and can be reused numerous times so once they start to fade and break up, they can be added to your food waste bin or compost heap. Soapnuts can be used to wash your dishes too – avoid mainstream dishwasher detergents that often use chlorine as a bleaching agent – by placing a few soapnuts in the cutlery basket of your dishwasher before a cycle or use as liquid detergent.

Conventional dry cleaning is energy- and water-intensive, and uses toxic chemicals that are usually released into the water system, namely sodium hypochlorite bleach, which causes caustic burns and respiratory issues, and sometimes a neurotoxic solvent called PERC (perchloroethylene) that's a probable carcinogen. Some dry-cleaning companies have replaced PERC with GreenEarth, a liquid-silicone-based solvent, while some people might prefer hand-washing dry-clean-only garments or machine-washing them on a delicate cycle.

Breaking the cycle

Tiny threads of synthetic material, such as polyester, acrylic or nylon (all made of petroleum-based plastic materials) are shed from our clothes every time we wash our clothes. Up to 1.5 million of these microfibre particles can be released from new synthetic clothes during a laundry cycle, and that's down to the chemical and mechanical stress that the fabrics experience during each wash. Synthetic clothes are the main source of microplastics that end up in our oceans, where they are impossible to retrieve.

Marine scientists at the University of Plymouth, UK, used a mesh to capture the fibres entering wastewater to test six fibre-catching devices that reduce the amount of microfibres that leave our washing machines.[73] The Guppyfriend mesh washing bag reduced the number of microfibres released by 54 per cent, while one of the three external filters tested cut down the quantity of microfibres being released by 78 per cent. The Cora Ball laundry ball filters out some microfibres. PlanetCare's microfibre filter can be quickly installed into any household washing machine and used cartridges can be returned and recycled. However, normal wear and tear when wearing clothes is just as significant a source of microplastics as release from laundering, so it's a good idea to opt for natural fibres, such as organic cotton and wool, as much as possible.

Professor Richard Thompson OBE, head of the Plymouth-based International Marine Litter Research Unit, believes that scientists need to collaborate with industry and policymakers to ensure improvements are made from the design stage of fabrics too: 'Some of the devices we tested can undoubtedly reduce the fibres generated through the laundry process, but perhaps the most overarching change would be to design garments to last longer and shed less fibres in the first place.' This philosophy is the cornerstone of a more circular and less toxic fashion industry, as we discover next.

TOXIC-FREE TAKEAWAYS

- In some countries, chemicals such as PFASs and lead end up in drinking water, so much better regulation is needed at source.
- Food labels don't tell the full picture. Everything from the history of the farmland to the packaging, processing and method of cooking affects the levels of toxic chemicals in what we eat.
- Homes don't need to be sterile; bleach is everywhere, but it's really not necessary. Good old soap and water is more than sufficient.
- Culturally, clean is associated with fragrances, but those synthetic smells give a false sense of security.
- Dish the dirt on surfactants and opt for plant-based ones that will more easily biodegrade in the environment.
- From cookware to cleaning, greener alternative solutions are already available.

6

In the Living Room

We spend most of our lives inside, so the air that we breathe in our homes really does matter. From building materials to the textiles and decor within our four walls, our indoor air can be bombarded by pollutants every time we light a candle or sit on the sofa. From printing inks to pesticides and cleaning agents, non-transport products now make up half the emitted VOCs in urban areas, so it's crucial that these sources of VOCs are tackled.

Although indoor air pollution has been generally overlooked until recently, it's nothing new. As early as 1500 BCE, the Egyptians were aware that the cutting of stones produced silica dust that could cause respiratory disease. Now, as we look towards a low-carbon society, it's clear that many of the solutions are inextricably linked to a less toxic future. Replacing the burning of fossil fuels with clean, renewable energy results in less polluted air, on an industrial scale as much as within our homes.

Indoor air pollution is a major public-health issue. Short-term impacts from exposure to toxic chemicals in the air include headaches, inflammation of the eyes, nose and throat or exacerbated respiratory conditions such as asthma. Longer-term health impacts could include impaired cognition, stroke or the onset of cardiovascular disease. But it's something that we can all do so much about because we have more control over the air chemicals generated inside our own homes than the quality of air we might breathe as we walk through a bustling city centre. The tricky bit is that harmful chemicals often come from invisible sources and build up in the air or in the dust that accumulates on surfaces. Some chemicals might be present at high levels, such as after lighting a wood-burning stove, while others might be present in low levels all the time, including VOCs that are off-gassed from furniture.

VOCs: The inside story

VOCs are emitted by a huge range of sources in the living room, from carpet cleaners, furniture polish, paints and plywood furniture to office equipment and air fresheners. While not all synthetic fragrances are harmful, their ingredients are usually not listed in full and we might not be told enough about what's in them. Scientists at Exeter University, UK, advise that commonly used household products should carry a warning that they increase the risk of asthma, as well as a reminder that people should ventilate their homes while using them.[74] The findings show that the risk of asthma increased by 40 per cent for people exposed to five VOCs and that even in people without asthma, wheezing attacks were more likely following exposure to products containing benzene, a carcinogenic solvent present in many household products.

Indoors VOCs don't dissipate quickly like they would do outside, and these chemicals – substances such as ethyl acetate and toluene – can be inhaled, ingested or absorbed through the skin. Concentrations of many VOCs are consistently higher indoors than outdoors according to the United States Environmental Protection Agency (EPA),[75] but in most parts of the world, legislation doesn't yet govern levels of VOCs that are considered to be safe in our home.

In urban areas, household cleaners and personal care products together with pesticides, adhesives, inks and coatings account for more VOC emissions than cars.[76]

As with all household products, what happens once they get washed away really counts. By using a less toxic carpet cleaner or stain remover, you'll end up washing fewer harmful chemicals down the drain. Instead of a polish spray for wooden furniture that can release VOCs and aggravate allergies, choose a plant-based polish in a bar form, or simply dust with a damp cloth. And be aware that air fresheners, plug-ins and deodorizers don't actually clean the air. They are a concentrated source of indoor air contaminants containing phthalates as well as chemical fragrances to overpower odours. Fragrances alone can contain hundreds of chemical ingredients, but fewer than 10 per cent of air freshener ingredients are usually disclosed to consumers on labels.[77] So, instead of synthetic air fresheners, try avoiding fragranced products in the home for a week. Use lemon juice as a cleaner if you're keen to have a natural scent while washing down surfaces. Bamboo-charcoal-based air-purifying bags like those produced by Moso Natural neutralize odours, attract harmful pollutants and absorb excess moisture from the air for up to two years. Think about any candles that you burn too – most are made from petroleum wax and they're often scented. If you light a candle, choose a fragrance-free one made with natural waxes such as soy, beeswax or coconut that will give a cleaner burn.

Emissions labelling on household products and building materials has a long way to go. In France, a traffic-light labelling system indicates the level of indoor air pollution that results from using a product and shows when ventilation during use is recommended. In Denmark, furniture that doesn't emit VOCs is sold with the Indoor Climate Label. Wouldn't it be helpful if these schemes were more widespread? Perhaps a new clean-air rating for homes could be developed, just like the energy-performance certificates that homeowners in some countries must provide before selling a home, so it's much easier to assess the air quality inside a particular building.

Sit down and put your feet up

From sofas and footstools to coffee tables and bookcases, furniture doesn't usually come with a materials list, so it can be hard work deciphering what it's actually made with. Plastic furniture can be manufactured using phthalates and BPA. The polyurethane foam used to make some furniture, sofas and playmats contains highly combustible materials derived from fossil fuels, so flame retardants are added to many household furnishings and these can have serious health impacts, especially for children.

TOXIC-FREE FURNITURE TIPS

- Opt for solid wood, preferably sanded without any paint or varnish or with VOC-free paint or natural waxes.
- Consider how a product is assembled. Glues and adhesives are often used to make upholstery and furniture, so sewn fabrics or wood that has been fixed together using tongue and groove is a better alternative.
- Avoid additional surface coatings on metal furniture wherever possible.
- Choose natural fibres – organic cotton and wool fabrics or jute and seagrass for durable rugs. OEKO-TEX certified textiles are made with limited formaldehyde.
- Buy second-hand furniture that is much less likely to off-gas VOCs than new items.
- Look for upholstered furniture that has been produced without flame retardants if available, depending on national flammability regulations.
- Avoid additional waterproofing and stain-resistant treatments and use water-based low-VOC paints.
- After painting your living room, wait a few days before introducing soft furnishings as they can absorb VOCs.

Medium-density fibreboard (MDF), once considered the DIY dream because it's so cheap and easy to cut and paint, is made of wood dust and scrap held together using a resin that contains formaldehyde, a known carcinogen. Formaldehyde off-gassing from furniture can have a direct effect on human health. VOCs can be released from the glues and adhesives that are applied during the manufacture of household furniture. Off-gassing reduces over time, so buying furniture second-hand reduces the risk. Alternatively, unwrap new furniture and let it air in the garage for at least a few days before installing inside your home, or phone the manufacturer a few weeks ahead of delivery and ask them to open the packaging to allow your product to sit in the open air of the warehouse.

Burn, baby, burn?

While there's something extremely cosy and comforting about lighting a fire and curling up on the sofa, burning wood indoors isn't a particularly healthy thing to do. Smoke from open fires and wood-burning stoves can contain huge volumes of PM2.5 particulates, black soot, benzene, formaldehyde and carbon monoxide. Depending on the type of fuel you burn and what you burn it in, breathing in air that's polluted with wood smoke can be harmful. In low-to-middle-income countries, 'dirty' fuels such as peat, crop waste, manure or coal are often still burned indoors and used to heat homes as well as cook food, thus contributing massively to indoor air pollution in homes with often very little ventilation.

Burning dry wood isn't as dirty as wet wood; coal is even worse than that. Domestic burning of wood and coal can produce more PM2.5 than road transport, despite only a relatively small proportion of people doing it, often in urban areas where air pollution is already a problem. As well as the tiny particles that can easily be inhaled once released, carcinogenic chemicals such as benzene, formaldehyde and polycyclic aromatic hydrocarbons are present in the mix. Solid-fuel open fires result in the most PM2.5 emissions, but wood-burning stoves produce a significant amount too. Wood burners triple the harmful pollutants found in indoor air, with pollution spiking every time the door of a wood-burning stove is opened to add more wood, as it effectively acts like an open fire and it can take an hour or two for these particles to dissipate.[78] If the door is kept shut, the pollutants head straight up the chimney, contaminating the air that your neighbours are breathing in.

If a room isn't well ventilated while a fire is lit, or if the chimney hasn't been swept regularly, poisonous and odourless carbon-monoxide fumes can build up. Carbon monoxide can result from faulty or unserviced gas fires, heating systems and cookers. Just as houses need smoke alarms, it's a good idea to put a carbon-monoxide detector alarm in the living room and kitchen.

Carbon-monoxide levels are raised by smoking indoors or through the use of faulty household appliances, especially when used in poorly ventilated rooms. In 2022, the new Ecodesign standard that becomes mandatory for new wood-burning stoves sold in the EU sets out efficiency criteria and particulate limits. These regulations ensure that the appliances produce less smoke, but the levels are still significantly higher than smoke generated by wood-pellet boilers or oil- and gas-fired boilers. So while they are an improvement, they aren't by any means the solution. The cleanest option is electric heating, which doesn't emit any particulates, and if your electricity supplier is 100 per cent renewable, even better. Your neighbours will thank you too.

TIPS FOR INDOOR FIRES

- Avoid burning wood indoors if anyone in the household is elderly, vulnerable to respiratory conditions, or if infants live in the home.
- If lighting a wood burner or open fire, make sure the room is well ventilated and that smoke goes up the chimney.
- Clean out your wood-burning stove regularly to remove any soot, ash or debris.
- Get your chimney swept every year.
- Remember that particulates will dissipate more quickly outside on windy days, but less so on calm, still days.
- Take note of any local air quality warnings and avoid burning indoor fires when air pollution levels are high.
- Burn briquettes, use wood pellets or clean, dry wood. Leave any wet wood to season and dry out for up to two years.
- Avoid burning rubbish or old, painted or treated wood that can release toxic fumes.
- Set the air flow on wood burners to high for fifteen minutes every time new logs are added.

Floored by dust

Dust in our homes comprises microscopic particles of dirt, pollen, bacteria, smoke, skin cells and hair. Many things inside our houses shed dust and contain traces of chemicals from the products we use on a daily basis. Our exposure to house dust happens in three possible ways. We can breathe it in, we can touch it and accidentally eat it if it gets onto our food or hands and then into our mouths. Some chemicals end up in our blood or even breast milk. This is a particular worry for babies and younger children who are more likely to transfer any dust that gets onto their hands into their mouths while crawling and playing on the floor. So, regular cleaning of surfaces with a damp cloth or mop, and handwashing, especially before mealtimes, helps to keep your contact with dust right down.

Chemical contaminants are just one aspect of our complex indoor environments. Everything from location to ventilation and the actual fabric of a building has an impact upon the levels of moisture, mould, noise and pests in a home. House dust gives a fascinating insight into the levels of certain types of chemicals present indoors that come from the materials making up our home, plus the products we use in it.

VOCs like formaldehyde won't be found in dust, but metals, flame retardants, forever chemicals like PFASs and semi-VOCs that are found in air and which are able to attach to particles and surfaces can all accumulate in dust. In homes near farms, pesticides might be more prevalent in dust, whereas lead might be found at higher concentrations in the dust within houses located near industrialized areas. Vinyl flooring can emit phthalates and homes with lots of vinyl flooring tend to have more phthalates in dust. Phenols and fragrances are commonly found in house dust, especially if scented candles and fragranced air freshener sprays are used regularly.

WHAT'S IN DUST?

- Phthalates from plastics and personal care products such as shampoo.
- Fragrances from household cleaners, scented candles and personal care products.
- Flame retardants from textiles, furniture, electricals and electronic goods.
- Microplastics from plastic packaging and other items.
- Phenols from paints and cleaning products.
- Pesticides from pets and garden use, or nearby farmland.
- PFASs used in non-stick coatings and fabrics.

Families with young children and endless foam-padded baby gear might have higher levels of flame retardants in house dust, and so replacing old upholstered furniture with flame-retardant-free furniture can dramatically reduce the levels of PBDE (polybrominated diphenyl ethers) flame retardants in your house dust. To save unnecessary waste, look online for foam suppliers in your area who could replace just the foam inside your furniture for a healthier option. Just always make sure old foam furniture is disposed of responsibly, preferably via your local tip or recycling centre, rather than ending up in the second-hand market where it could unknowingly add to someone else's toxic load.

The indoor environment is certainly not homogenous – it varies between homes, from day to day and from country to country. House dust from hundreds of homes in twelve countries, including Greece, India, Japan, Pakistan and Saudi Arabia, was found to contain high concentrations of PET (polyethylene terephthalate), the plastic commonly used to make drinks bottles. The greatest amount was found in a dust sample from China, the largest producer of PETs globally.[79] Levels of PFAS forever chemicals in house dust were compared for eight countries including Sweden, the Faroe Islands, Australia and Canada. The quantities in the dust samples mirror the usage of PFAS-containing household products in those places. Samples from Japan, a wealthy and highly developed country, consistently showed the highest concentrations of PFASs in house dust, whereas the lowest levels were found in Nepal, a far less densely populated country with a much less industrialized economy.[80]

When Veena Singla, senior scientist at the Natural Resources Defense Council, looked at house dust specifically for chemicals found in the commercial consumer products in current use, she was surprised by her findings. Things like legacy contaminants, pesticides and pharmaceuticals were excluded from her search, yet she still found forty-five toxic chemicals, ten of which were in more than 90 per cent of the house dust tested. Compared to the EPA's health-risk-based soil-screening levels of chemicals, the quantities of phthalates and flame retardants found in house dust were really high – even some of the averages exceeded recommended amounts in soil. But there isn't yet an accepted healthy, safe level for chemicals in house dust. Generally, it's forgotten about. Plus, in soil, chemicals get washed away by the rain or broken down by microbes. Inside our homes, we need to actively remove it.

House dust accumulates on the clothes we wear, and when they get washed, things like flame retardants end up in the waste laundry water. These chemical contaminants aren't always specifically targeted in the wastewater treatment process. Although house dust might seem insignificant, it's a reminder that our indoor environments are a crucial source of chemical exposure. As with any persistent chemicals, there is no 'away', so they'll just keep circulating unless specific action is taken to remove them. Ultimately, it's best to avoid them in the first place.

So which is best – hard floors or carpet? Hard floors might be perceived to be more cleanable, but really they just make it easier to spot the dirt, so it's more likely you'll clean them. Instead of sweeping hard floors, it's best to clean with a wet mop and hoover carpet using a vacuum with a HEPA (high-efficiency particulate air) filter, which traps dust and PM2.5 that accumulate in the carpet fibres. HEPA filters are used in air purifiers and can significantly improve indoor air quality. In Ulaanbaatar, Mongolia, one of the most polluted capital cities in the world where coal-burning stoves are used throughout the coldest months, severe air pollution has been linked with high incidences of respiratory diseases and miscarriage. Here, the use of HEPA filters has been shown to reduce blood cadmium levels in non-smoking pregnant women as well as PM2.5 and second-hand smoke.[81] Simple interventions like these really can improve quality of life.

Unlike synthetic carpets, wool carpets can help to purify the air as the fibres naturally absorb common VOCs such as formaldehyde, sulphur dioxides and nitrogen oxides for up to thirty years.[82] Wool only absorbs VOCs from the air it's in direct contact with – so wool furniture, floorings and wall coverings can be beneficial – but if it's used as insulation within a wall cavity, it won't do this. Wool naturally stabilizes the humidity in your living room because it can absorb large quantities of moisture, minimizing the risk of developing mould. Wool is naturally flame-resistant, so chemical flame retardants don't need to be added. Many carpet cleaners include chlorine bleaching agents which are toxic and stay embedded on the carpet fibres after being applied, but wool-friendly cleaning products tend to have fewer harsh chemical ingredients, so look for the international WoolSafe guarantee.

Watch out for those stain-resistant, waterproof claims on synthetic carpets and rugs too, because that indicates that PFAS chemicals and PFAS-containing Scotchgard treatments have most likely been applied. In 2019, the world's largest home-improvement retailer, The Home Depot, banned PFAS chemicals in the carpets and rugs it sells across the US, Canada and Mexico, and more carpet manufacturers are gradually starting to phase out PFASs.

TIPS FOR FRESHER AIR

- Ventilate your home well every day.
- Avoid air fresheners and buy fragrance-free products. A few drops of vanilla extract can be used instead of air fresheners and scented candles, while a bag of dry coffee beans in a thin muslin cloth can remove odours.
- Avoid smoking or the use of e-cigarettes indoors.
- Spray any cleaning products close to the surface and make sure they get wiped off.
- Reduce the number of different spray products you use in your home and choose ones with fewer ingredients.
- Minimize the presence of VOC-emitting products in the home. Opt for low-VOC alternatives where possible.

What's in my hoover bag?

'House dust is like the soil of the indoor environment,' according to Stuart Harrad, professor of environmental chemistry at the University of Birmingham, UK. 'Settled floor dust acts as a sponge for all the chemicals flying around in the air that end up depositing onto surfaces so it gives us an insight into the profile of chemicals present within a home.' I sent Harrad a sample of dust from my hoover bag which he analysed for twenty-four flame retardants, some of which are banned, and the results were intriguing. 'Most of the concentrations we found are lower than what we might typically see across the UK. We'd expect to see the highest concentration of one called BDE209 [decabromobiphenyl ether] that has only recently been banned in the UK, but yours was very low,' Harrad told me. He found high concentrations of two PBDEs that are rarely found in UK samples. One is most commonly found in North America, one is often seen in samples from the Middle East. I do use the hoover to clean the car so one possible source could be the children's car seat or perhaps a rug bought on holiday.

So should I worry? Harrad doesn't think these results are of immediate concern: 'Even adults will accidentally ingest something like 20 mg of dust per day, that figure is higher for young children. But even at these concentrations, I'd expect your exposure to dust to be well within the safe limits set by the US EPA.' These tests have spurred me on to vacuum every room more frequently – Harrad suggests weekly. Plus I use a damp cloth to wipe down surfaces, including TVs, routers and computers because when dust settles on these items, flame retardants can transfer out from these products. The trick is to remove it before it accumulates and before people can be exposed to it.

WHAT'S A PASSIVHAUS?

The Passivhaus standard for energy-efficiency is widely recognized as the ultimate benchmark in eco-friendly living. Literally meaning 'passive house', these homes are well insulated, well ventilated and constructed to eliminate all draughts, and there is no need for any central heating, hence the name 'passive'. Far from a sealed box, the airtight yet breathable fabric of the walls keeps a Passivhaus cool in summer and warm in winter. Fresh, dehumidified, filtered air circulates around the building thanks to a mechanical ventilation with a heat-recovery system that allows for the flow of fresh air without letting any heat escape and dramatically increases energy-efficiency. With the appropriate filters installed, this heat-exchange technology takes fresh air from outside and filters out any airborne germs, dust, particulates and pollen. The Passivhaus standard has grown in popularity across Europe and, most recently, in South East Asia. In China, where outdoor air pollution can be a big problem, thirty-seven buildings in the new city of Gaobeidian are being developed entirely to the Passivhaus standard, including high rises and villas. This will be the world's biggest ever Passivhaus project and is due to be completed by 2023.

Snakes and spiders: an urban myth

From snake plants and spider plants to peace lilies, there's a lot of hype about the importance of having houseplants in your living spaces. But can plant power really purify the air inside our homes? NASA looked into this back in 1989. They tested how plants like peace lilies, gerbera daisies and bamboo palm could absorb chemicals such as benzene and formaldehyde within sealed chambers in a lab. They found that the plant roots and the soil were doing most of the filtering, not the leaves. Scientists have since discovered that it could actually be the microbes in the soil that are involved in this process, not the plants themselves.[83]

Over the years, so many companies have capitalized on this plant power belief as a clever marketing ploy, but unfortunately the NASA study doesn't realistically translate to real life. We don't live in an enclosed, controlled environment and we'd need an incredible number of houseplants or vast green walls in our homes to even come close to replicating a similar scenario.

Houseplants won't do any harm and they are considered a key part of the biophilic or nature-loving principles used in interior design that incorporate elements such as plants, light and water into a room. More living walls packed with mosses, ivy and ferns are popping up on and inside buildings to try to clean up pollutants in urban areas, but it's hard to quantify exactly how much difference they are making.

Choosing healthier furnishing materials for inside your home will dramatically reduce the amounts of chemicals such as flame retardants and stain-repellent PFASs in your house dust.[84] Aside from not introducing toxic chemicals into your home in the first place, one of the most important things that you can do to reduce indoor air chemicals is to have adequate ventilation. Where we install windows, how big they might be and how easy it is to access fresh air really matter, and there's a growing movement of architects designing healthier buildings. Assuming the outdoor air isn't more polluted than inside your home, simply open windows and doors. If you live in a city centre, by a main road or near an industrial area, there might be a need for a more technological solution. So, if you're seriously concerned about indoor air quality, it makes sense to invest in a way to mechanically ventilate and purify your indoor environment. Air purifiers are best placed in the room where you spend most of your time, probably the bedroom, and the room where most pollutants are found – this may be the kitchen or living space depending on your situation.

I have been experimenting with a BlueAir HEPA purifier in our open plan kitchen living room. Even boiling the kettle can shift the light from a pure blue to green, perhaps because the steam is moving dust and other contaminants around. The most dramatic orange and red spikes on the chart represented particulates and VOCs released while making toast, roasting food in the oven and accidentally overheating oil in a pan on the hob. So despite not smoking cigarettes and not having an open fire or wood-burning stove, it has really highlighted the need to minimize preheating, ventilate during cooking and clean away dust more regularly.

The e-waste mountain

There's a lot lurking in most homes that won't ever be used again, from tangled cables and old remote controls to redundant VHS players and mismatched phone chargers. An estimated 50 million tonnes of e-waste is produced globally every year, weighing more than all of the commercial airliners ever made.

Only 20 per cent is formally recycled.[85] Rapid growth of the global e-waste mountain is driven by increasing consumption of electrical and electronic equipment (EEE), short life cycles and limited options for repair. Once manufactured and used, an item that contains precious metals and other mined materials often gets discarded and ends up in landfill. E-waste is an 'urban mine' with so much potential for recovery of valuable metals that could reduce the need for us to use virgin materials. At the moment, it's an untapped resource.

Sixty-nine different elements can be found in EEE, from gold, silver and platinum to copper, iron and aluminium. EEE also contains hazardous substances, such as carcinogenic heavy metals like lead, mercury and cadmium, plus CFCs (chlorofluorocarbons) and HCFCs (hydrochlorofluorocarbons) used in older refrigerators and air-conditioning units, or the brominated flame retardants used in the outer casings of computers, cables and wires. These all pose a serious challenge to environmentally friendly and healthy disposal. Around the world, children play or work in informal, unregulated recycling sites and there's a greater risk to pregnant women working as recyclers who might experience adverse health effects, from birth defects to respiratory problems and cancer. Currently, a lot of collected e-waste ends up being sent to developing countries where, once high-value metals have been removed, often the plastic components get burnt, releasing toxic dioxins in the process.

Recycling needs to be better managed globally, but solutions lie at the very start of the supply chain, before anything is even manufactured. Often metals are glued together and difficult to extract, so products need to be designed to be disassembled efficiently. PVC is used as coatings for computer cables and wires, and in the past flame retardants such as PBDEs were used in electrical items for

CLOSING E-WASTE CYCLES

The PREVENT Waste Alliance (prevent-waste.net) is piloting new ways to close e-waste cycles and avoid toxic waste pollution. In Africa, it is supporting the Tanzanian government to better reduce and control imports of waste. It's designing and implementing the first e-waste collection scheme in Ecuador as part of Quito's circular strategy. And in Nigeria, a waste compensation project is finding ways to provide financial incentives to recycle 20 tonnes of lithium-ion batteries and flat-panel display screens in the hope that effective e-waste recycling can be scaled up.

safety reasons, so older electricals can be harder to recycle. Batteries should be handled very carefully to ensure leaching doesn't occur because they aren't designed to be squashed. Items have to be designed with end of life in mind to ensure that they can be easily disassembled and elements reused. It could be so much more energy-efficient and streamlined.

From phones and laptops to cables and speakers, anything with a plug or a battery can be recycled. Recycling of e-waste needs to become more of a mainstream habit and convenience is key, but there's a disconnect in the system and the associated language can be confusing. When you first buy an appliance, it seems shiny and attractive, but the moment you dispose of it, it's usually classified as hazardous waste.

Some large retailers, including supermarkets, are starting to offer electrical recycling drop-off points, so that customers can return unused items for free. Perhaps one day, e-waste recycling will be as commonplace as going to the bottle bank.

Do you really need that upgrade? There's so much in-built obsolescence that doesn't make logical sense, especially when it comes to tech. Many vintage electricals and design classics can be repaired and reused, and there's nothing wrong with second-hand IT equipment if you purchase from a reputable supplier such as Reconome or Circular Computing, which has a useful sustainable IT calculator to assess the impact you could have.

E-waste is the fastest growing domestic waste stream in the world.

Apple's disassembly robot, Daisy, can take apart up to 200 used iPhones per hour and recover valuable metals and other materials. Fourteen minerals, including lithium, are being extracted and recycled when iPhones are sent back to the retailer via the Apple trade-in programme.

But recycling alone still wouldn't meet the excessive consumer demands of an ever-growing population. We can mine less by digging deep and consuming less. Having fewer mobile phones, laptops and tablets, or replacing them less often with refurbished alternatives, makes a big difference and helps to push back against premature obsolescence and unnecessarily short lifespans of electronics. Proposed 'right to

TIPS FOR TOXIC-FREE METALS

- Quiz manufacturers and retailers about where their minerals and metals are sourced from.
- Join the repair revolution. Search ifixit.com to find out how your gadget could be repaired.
- Recycle your e-waste and value key items as long-term investments rather than being disposable. Buy reconditioned mobile phones direct from your provider or from reboxed.co in Europe.
- Consider buying an ethical phone that supports responsible mining. Available across Europe, Fairphone is made with recycled plastics, conflict-free minerals and Fairtrade gold, plus it's designed to last with seven easily repairable modules.
- Look for TCO Certified (tcocertified.com), the leading sustainability certification for IT products that have met certain social and environmental standards throughout the product's life cycle.
- Handle batteries with care – they aren't designed to be squashed or punctured and chemicals can leach out of them – and recycle them at a designated battery-collection point.
- When buying jewellery, hunt for vintage or antique pieces of gold and silver instead of buying brand new.
- Buy gold with the Fairmined assurance label that's sourced from responsible artisanal and small-scale mining organizations.

repair' legislation will make software publicly available in order to enable someone to fix it, rather than inhibiting the ability to make do and mend.

From electricals to toys, furniture to fabrics, lots of hidden toxic chemicals can linger in the living room, but with a few simple changes our exposure to them can be minimized. Whether we spend a lot of time working, resting or playing, our communal spaces are fairly full of stuff. How we decorate and furnish our living space can have huge knock-on effects for the people in that home and the surrounding environment. Choosing second-hand furniture, hard flooring and electric heating can all reduce risks of exposure to certain chemicals. Avoidance of VOCs can be a breath of fresh air and decluttering is a key step to keeping your living room free from dust and invisible toxic chemicals, making it easier to concentrate, socialize or just relax.

TOXIC-FREE TAKEAWAYS

- Indoor air quality can be controlled to some extent through better ventilation combined with fragrance-free and low-VOC choices.
- The profile of toxic chemicals in a home varies over time, by location, and depends on the building type and lifestyle choices.
- As cosy as a wood-burning stove or open fire might seem, the associated PM2.5 levels are quite shocking.
- So many different chemicals accumulate in house dust, which can be more of a hazard to families with young children and pets.
- Electrical waste is full of useful resources, including precious metals. Repairing, recycling and reusing them reduces the demand to mine more materials from the Earth's crust.

7

In the Bedroom

A third of our life is spent sleeping or attempting to do so. Of all the rooms in our home, the bedroom is where we spend most of our time, albeit with our eyes closed. From the wardrobes we buy to the linen we sleep underneath and all the clothes, shoes and accessories we wear, there's a huge array of fabrics, textiles and materials in every bedroom, each with their own impacts. Back in the 1970s, environmental scientist Arlene Blum sounded the alarm on a brominated flame retardant known as tris that was being used in some children's sleepwear. A likely carcinogen, this flame retardant was showing up in children's urine after wearing tris-treated pyjamas and her work led to the use of these chemicals being banned for this purpose.[86]

But plenty of other toxic chemicals could still be hidden in your bedroom, many of which are odourless, so not always noticeable. So, how can we reduce health risks and pollution from chemicals used to make our furniture, foam and fashion?

Between the sheets

When it comes to what we sleep on, it's worth finding out what your mattress has been made of because some can take a decade to stop emitting fumes. Due to the length of time we spend in our bedrooms, off-gassing can really affect indoor air quality and our bodies while sleeping. Our body heat increases the temperature in our beds, so more VOCs are released during the night from conventional polyurethane mattresses. Some of the toxic chemicals, including acetaldehyde, formaldehyde and benzene, can reach levels that would be of particular concern to young children.

Most mattresses tend to be made from polyurethane foam, polyester, adhesives and then treated with flame retardants, and stain-resistant or antibacterial chemicals. Because polyurethane foam is derived from fossil fuels, it's highly combustible. To combat this fire risk, persistent flame retardants are added – these chemicals include organophosphates or halogenated flame retardants, which are banned by the Stockholm Convention in Europe but still permitted by some US states and other countries.

Instead of buying a new mattress, invest in a decent topper that is either made of 100 per cent natural materials, such as wool, or latex, allergies permitting. Wool works well as a natural temperature regulator and latex is naturally flame-retardant, anti-fungal, antibacterial and resistant to dust mites. A wool futon mattress is a good budget option and some cotton mattresses are made with wool woven around them to act as a natural flame retardant.

A few companies make beds and mattresses that are free from synthetic materials or chemical treatments such as organophosphate fire retardants. The UK-based Naturalmat specializes in bed products made from only natural, sustainably sourced and sometimes upcycled materials, including organic lambswool, organic coconut fibre, mohair, horsehair, bamboo and cashmere. They source waste offcuts of cotton denim from the fashion industry for padding and they use Rainforest Alliance Certified natural latex foam made from the sap of Hevea (rubber) trees that's filled with air bubbles. Mattresses that have been handmade and sewn won't have been glued together with adhesives. Wool slows down the burn rate of the fabric, and unlike organophosphates, which create a toxic flame when set on fire, wool doesn't give off any toxic gases. Once conventional mattresses get sent to landfill, toxic chemicals can leach out into the waterways, so look for companies like Naturalmat that offer a mattress recycling service.

While it doesn't make any environmental sense to replace everything in your bedroom, if you do need to buy new bedding, consider unbleached natural materials that haven't been treated with as many synthetic chemicals and you can't go far wrong. When buying duvets, pillows and bed linen, look for certified organic and cotton with the GOTS (Global Organic Textile Standard) label. Bedding that claims to be easy-care or anti-wrinkle will have had extra chemicals added at the finishing stage, such as PFASs for stain repellency or antimicrobial nanosilver. Waterproof mattress covers are often made with PVC and phthalates.

If you choose feather fillings, ask the manufacturer about their practices and choose cruelty-free feather down. Suppliers should be using feathers that are a by-product of duck and geese farming, rather than live plucking. Textile Exchange, the global non-profit that works with all sectors within the textile supply chain, lists certified suppliers that carry the Responsible Down Standard label (responsibledown.org) that ensures that any down or feathers come from animals that have been treated humanely.

Chemicals in the closet

Most clothes hanging in your wardrobe might seem fairly innocent at first glance, but every garment on a hanger is a snapshot of a convoluted supply chain. The sheer scale of mass production and the vast, global nature of fast-fashion supply chains is quite shocking when it comes to toxic chemicals. Yes, they're safe to wear now, but the fast-fashion industry is toxic to its workers. In Europe alone, 15,000 different chemical formulations are used in textiles manufacturing on a regular basis.[87] But so much gets wasted in the process because it takes between 10 to 100 per cent of the weight of the fabric in chemicals to produce that material.[88] In 2019, approximately 111 million tonnes of textile fibres were produced – and less than 1 per cent gets recycled into new clothing.[89]

The exact combination of the chemicals used in the fashion industry varies enormously depending on the garment. Pesticides are used to grow cotton, waterproof synthetics are made with PFASs, and most clothes have been dyed with pigments and mordants to fix the colours. While many of these chemicals get washed out during the manufacturing process, some are added as finishing agents at the end to give 'added benefits'.

Fashion accounts for 10 per cent of global carbon emissions – that's more than shipping and aviation combined.[90]

Perhaps the biggest culprit is conventional cotton. Cotton makes up 90 per cent of all the natural fibres used in textiles, but it's a very thirsty, toxic crop that involves using a lot of land. It takes 2,700 litres of water to grow enough cotton to make just one T-shirt. That's enough for one person to drink for 900 days.[91] Huge volumes of pesticides are sprayed on conventional cotton crops and synthetic chemicals are used throughout the wet processes of dyeing, bleaching, printing and finishing during manufacture. Denim jeans are a classic example. The majority of us own more than one pair of jeans, most of which get made in the US, Mexico and China. It takes hundreds of gallons of water to make just one pair of jeans, and synthetic indigo, the dye used to make denim jeans blue, leaches out from factories into the waterways. This continues when we wash our clothes at home as they fade.

In Uzbekistan, Central Asia, conventional cotton farming has transformed the world's fourth largest lake, the Aral Sea, into a salty desert. In August 2014, images from NASA revealed that this saltwater lake had almost dried up. Once home to a thriving fishing industry, the freshwater rivers that fed into the lake had increasingly been diverted since the 1950s to irrigate the surrounding 1.47 million hectares of agricultural land used to grow cotton. Now, poisonous dust storms carry 43 million tonnes of toxic salts and sand from the dried-out sea floor through the air. This dust is contaminated with pesticide residues and other toxic pollutants. Dioxins and polychlorinated biphenyls (PCBs) have been found in fish, sheep, milk, eggs and other foods, while carrots and onions, vegetables that are important to the local diet, contain high pesticide concentrations. Scientists have warned that living near the Aral Sea area has detrimental consequences for fertility in men and women.

The future of this toxic lake hinges on the economic and political will to invest in solutions. In the north, dams have been built in an attempt to recover the water. In the south, thousands of salt-tolerant shrubs, grasses and trees have been planted to try to transform the now so-called Aralkum Desert and help to reduce the toxic sandstorms and dust storms.

The Better Cotton Initiative (BCI) aims to reduce the environmental impacts of cotton cultivation. But textiles labelled as BCI simply indicate that the company supports the production of BCI cotton – it doesn't mean that the garment has been made using BCI cotton, so it's no guarantee that textiles will be free from pesticide residues. Certified organic cotton is better for the soil and rivers, uses less water and energy in its production, and supports a more ethical economy. It's safer for the farmers and women harvesting the crops, and is better for your skin too.

Sometimes viscose / rayon is promoted as a sustainable alternative to conventional cotton or oil-derived polyester because it's based on cellulose, a plant material. It's derived from the wood pulp of fast-growing trees such as eucalyptus, beech and pine, and its production contributes to the rapid depletion of forests with 70 per cent of trees being wasted. Every year, it is estimated that 200 million trees are logged to make viscose/rayon. Crucially, this is a chemically intensive process. Fibres are treated with a solvent called carbon disulphide and toxic fumes are emitted in the process, affecting those who work in the textiles industry and people living near the viscose factories. This has been known about by the UN Environmental Programme and WHO since 1979, and carbon disulphide has been linked to higher levels of coronary heart disease, skin conditions, cancers and birth defects. Alternatives are being developed, though. EcoVero is a new EU Ecolabel-approved fabric alternative to viscose that is

TOXIC-FREE BEDROOM TIPS

- **Ventilate:** use a fan or open your bedroom window every day and create fresh air flow.
- **Sleep:** choose organic cotton bedding where possible. Avoid Teflon-coated or waterproof mattress covers – wool is naturally water-resistant. Buy mattresses without flame retardants. If you already own a mattress made with polyurethane foam, consider investing in a new mattress topper made from natural materials.
- **Furnish:** let new plywood bed frames and cupboards 'breathe' to allow for off-gassing before use; alternatively, buy second-hand items.
- **Decorate:** resist the temptation to paint your baby's nursery just before they are born in order to limit exposure to harmful VOCs. Paint bedroom walls with low-VOC paint such as that from Little Greene. Choose fabric curtains or wood-slatted blinds rather than vinyl blinds or roller shades made from PVC.
- **Clean:** wash bed linen with fragrance-free laundry detergents. Opt for easy-to-clean hard floors with a rug made from natural fibres or choose carpets that don't contain flame retardants or stain-resistant treatments.
- **Play:** in children's bedrooms, avoid very old second-hand toys which might contain lead paint and opt for washable soft toys rather than plastic ones.

produced using renewable wood sources, grown in responsibly managed forests and is fully biodegradable within three months. Plus, the Canadian NGO Canopy is developing solutions that focus more on the recycling of wood pulp and the use of straw left over from grain harvests as a viable fibre source for clothing and fabric.

The manufacture of leather for belts, jackets and handbags can be notoriously toxic to factory workers and the environment. Tanning is a process that prevents animal skins or hides from decomposing by treating them with chemicals to dry them out and create leather. It can involve formaldehyde, acids, dyes and salts of the metal chromium, which speeds up the process. Chromium salts in leather dust have been linked to health problems in tannery workers such as rashes, skin bleaching, nose bleeds and respiratory problems. In 2015, the volunteer doctors of Médecins Sans Frontières helped to treat people with chronic skin conditions and lung disease who had been affected by toxic chemicals used by tanneries in Bangladesh; this was the first time this charity has intervened in an area for reasons other than natural disasters or war. Some of the workers were children as young as eight.[92]

Chrome tanning is quick – it can take just a week – and it's relatively cheap, plus it makes leather more flexible than other tanning methods, but only a fraction of the chemicals used in the tanning process are actually absorbed by the leather. So where does the remainder go? Closed-loop systems ensure that no waste escapes into the environment, and that water and other resources are reused. But if that waste isn't captured and treated, it's discharged into the rivers, streets, soil and air, contaminating crops and drinking water. In Kanpur, India, more than 300 tanneries make leather accessories such as belts and handbags for the US and European markets, then dump toxic wastewater contaminated with chromium into the waterways and eventually the River Ganges. In 2013, scientists described the tannery industry effluent as 'brown, turbid and with an offensive odour'. Later that year, the government temporarily closed Kanpur's tanneries and regulation now requires chrome recovery plants to be fitted to all tanneries.

As with any chemical, toxicity comes down to its application, and some tanneries use chrome tanning with minimum impact and almost zero waste. So, if you're buying any animal leather, purchase from a reputable brand that is open about its supply chain and production process. The Leather Working Group has certified more than 750 leather manufacturers in forty-four countries around the world, and the ones that adhere to the highest gold rating demonstrate greater efforts to reduce environmental impacts. Look out for vegetable-tanned leather, which is treated with plant extracts or tannins from tree bark, berries or leaves to colour and preserve the hides. This process produces less harmful waste and a softer, more biodegradable leather than those made using chromium, but because it takes longer, it tends to be more expensive.

Buying second-hand or vintage leather is a great option, but of course many people disagree with the ethics of leather altogether from an animal welfare perspective and there are lots of vegan leathers now coming onto the market. Are these any better for the environment? Many so-called vegan leathers are made from petroleum products; effectively, some are just plastic that won't ever degrade. The most common types are polyvinyl chloride (PVC), which is made using chlorine-heavy processes and contains carcinogenic chemicals, and a plastic polymer called polyurethane (PU).

Some vegan alternatives are plant-based, but these materials are often mixed with a plastic element or PU coating to strengthen them because they tend not to be as strong or durable as animal leather. In Mexico, a company called Desserto makes cactus leather. As a crop, cacti don't need any irrigation, and no heavy metals, solvents, phthalates or PVC are used in cactus-leather production. VEGEA, an Italian company, transforms waste grape skins from the wine industry into grape leather, and in the Philippines, waste pineapple-leaf fibres are made into Piñatex. Other leathers are being developed from corn and cork, while MuSkin is made from the mycelium of fungi. Just avoid any heavy finishes or colourful coatings, and look at whether materials have been glued with more chemicals or sewn together. If they contain plastic, they won't be biodegradable or easily recyclable.

Fixing fashion

The global fashion industry is so vast, complex and fragmented, which ensures there's no easy or quick fix, but greater transparency across the board is absolutely crucial to reducing the impacts on ourselves and the planet. In 2013, British fashion designer and social entrepreneur Carry Somers founded Fashion Revolution, now the world's largest fashion activism movement, following the Rana Plaza factory collapse in Bangladesh which killed more than 1,100 people. Somers' vision is for a global fashion industry that values people over growth and profit, while cutting out the toxic environmental pollution: 'Typically for fabrics imported to the EU, a kilo of fabric would have used about 500 grams of chemicals, including dyes,' said Somers, who created the Fashion Transparency Index to hold brands to account in markets from luxury to high street, sportswear to denim. The index rates 250 of the world's biggest fashion brands according to how much information they disclose to customers about the chemicals they use to make our clothing. Now, she's pushing for legislation that requires suppliers to reveal details about the chemicals they use to make clothing.

As conscious consumers, we must keep asking questions about how our clothes are made. Most textiles are manufactured in China, which consumes 42 per cent of textiles chemicals used globally every year.[93] A new priority chemicals list is due to be released by China's Ministry of Ecology and Environment, which should restrict six types of phthalates in textiles products made for children under three years, but this really is only the tip of the chemical iceberg.

Senior scientist at Greenpeace Research Laboratories, Dr Kevin Brigden, has studied the toxic chemicals found in clothes and has noticed a major shift in awareness across the fashion industry over the last few years. In 2014, Brigden and his team of chemists tested a cross section of clothes for sale around the world and found an array of residues of chemicals on the fabrics.

As part of the Greenpeace report into hazardous chemicals in well-known brands of children's clothing, Brigden found surfactants called nonylphenol ethoxylates (NPEs) in products from ten of the twelve countries of manufacture, and 61 per cent of the eighty-two products tested contained these chemicals. After washing, NPEs in the environment are partially broken down into persistent nonylphenols (NPs) which act as endocrine disruptors. These are highly toxic to aquatic life and cause feminization of fish, plus they accumulate in sediments and build up in food chains.

Alternatives to NPEs are available, but the supply chains are long and often regulations are not consistent between continents. In the EU, the use of NPEs in the manufacturing process was banned to stop these chemicals being washed into the waterways and water treatment plants. However, the sale of clothing containing residues of these chemicals was not prohibited, so for some time these clothes could still be imported. When people washed these clothes, residues would continue to be released into the environment. Now, in the EU, the amount of NPE permitted on a product being sold is limited.

But there's no point in restricting the chemicals found in clothing if it just results in the manufacturer washing products more before exporting them. That only shifts the problem. Some clothes might be made with a whole range of nasty chemicals that get washed out of the clothing by the time it reaches the consumer, but those chemicals will have still ended up in the waterways near the garment factory.

Elimination is the best policy so that hazardous chemicals aren't used in the first place. Big brands have a powerful role to play in this. If a high-street fashion brand makes a conscious choice not to sell clothes made using chemicals of concern, there'd be enormous repercussions. This would influence the production facilities and trigger a more environmentally sustainable supply chain.

Detoxing clothes

So is it dangerous to wear clothing? There are some examples of chemicals in clothing doing harm to the wearer, though that's a rarity, Brigden told me. But the presence of those chemicals in our clothes indicates their use and potentially widespread release into the environment, both during the manufacturing process and then when being washed multiple times and later disposed of.

When Brigden compared the chemical profile of garments before and after the first wash, he found that results vary considerably depending on the particular chemical. Some persisted, whereas others such as formaldehydes and NPEs washed out more easily.[94] Up to 94 per cent of NPEs could be washed out by a first clean of the fabric, depending on the brand. But there are two sides to every spin cycle. This might seem healthier for our bodies, but these chemicals end up in our environment where they can continue to have knock-on effects. So the ultimate goal is to reduce the number of chemicals used in production and contain the impacts.

Greenpeace's work emphasizes that by focusing on chemicals used during clothing manufacture, schemes like its Detox My Fashion campaign are much stronger than ones that solely look at what is in that product at the point of sale. Since July 2011, Greenpeace has secured detox commitments from more than eighty global brands to eliminate hazardous chemicals, including Adidas, H&M

MOTHS IN THE CLOSET?

Mothballs contain pesticides – to avoid moth infestations store wool, cashmere and silk clothes that you don't wear regularly in airtight containers and never wear clothes straight after using mothballs; always wash them first. Instead of mothballs, you could try using cedar blocks as natural moth repellents. It helps to vacuum with a HEPA filter to hoover up all life stages of moths. If your clothes do get infested, pop them in a bag and freeze them, ideally for a week.

and Puma. With specific timelines and targets, brands have to be transparent about what they are doing and declare which chemicals are found in their wastewater. As a result, this has helped to change policies in Europe and Asia.

In response to Greenpeace's detox campaign, an industry initiative called the Zero Discharge of Hazardous Chemicals (ZDHC) developed the Manufacturing Restricted Substance List (MRSL), which highlights substances that have been banned by member companies from intentional use in their products. Wastewater guidelines have been set and 98 per cent of tested facilities in ZDHC's programme now have no detections of the eleven hazardous chemical groups identified by Greenpeace – priority chemicals including dyes and colourants which can bioaccumulate in the body over time and be highly toxic. That's a dramatic shift and the industry is now taking these issues more seriously. More collective action like this is needed. By combining initiatives like ZDHC with clearly defined regulations that impact the whole global supply chain, a new standard can be set for the entire fashion industry.

Fashion produces about one-fifth of industrial water pollution – from dyeing and textiles treatment – and a tenth of global carbon emissions.[95] Furthermore, an astonishing 79 trillion litres of water is used by the fashion industry every year.[96] Most textiles factories are in developing countries and the Natural Resources Defense Council has introduced its Clean by Design initiative – in partnership with some major fashion brands and designer Stella McCartney – to encourage best practice to save fuel, water and electricity while tracking water, steam and electricity use. The ten-step process aims to reduce the impacts of the most polluting aspects of the textile production process – dyeing (applying dyes or pigments to add colour) and finishing (applying chemical or physical surface treatments to give the desired aesthetics or technical performance). Almost all chemical discharges into the environment occur during these wet processes.

Step inside the shoe industry

According to World Footwear, in 2018 the global footwear industry produced 24.2 billion pairs of new shoes. That's enough to circle the world 300 times and 90 per cent of them won't ever get recycled. The harmful chemicals and toxic wastewater produced by the footwear manufacturing industry have huge repercussions on the environment and most shoes end up in landfill or incinerators. Scientists in Germany have found that shoe sole abrasion is the seventh largest contributor of microplastics to the environment at 109g per person every year.[97] PFAS chemicals are often added for durable water repellency, and on top of that, in the quest to produce vegan shoes, many manufacturers have turned to PVC, a synthetic plastic polymer that cannot be recycled and contains carcinogenic chemicals.

When KEEN Footwear realized that more than a hundred components of their shoes, from the laces to the stitching, were made with forever chemicals from the PFAS family, the company had a real wake-up call. They put a stop to almost 70 per cent of their PFAS usage immediately, just by ending non-essential use. For the remaining 30 per cent, they began a mission to find safer alternatives and work right back along the supply chain to make sure these chemicals weren't being added. It took four years for the company to achieve their goal and produce shoes free from PFASs, and now they're sharing their process so that others can eliminate PFASs too.

Sustainable brands like Ethletic use FSC-certified natural

rubber, Fairtrade-certified and organic cotton, nothing gets airfreighted and repair options are available. Some designers are embracing the circular philosophy even further and producing shoes made with materials that can be reused or eventually composted. 'It's time for a zero-waste revolution in the fashion industry,' said WAES Footwear co-founder Ed Temperley, who believes that recycling won't solve the plastic pollution problem and has created the world's only 100 per cent plastic-free sneakers. Every pair is organic and biodegradable. Manufactured in Portugal, WAES Footwear shoe soles are made from sustainably sourced sap from rubber trees grown in plantations in South East Asia.

Naturally colourful

Since 1856, when chemist William Perkin invented the first artificial colour 'mauvine' from coal tar, vibrant dyes with a petrochemical base have become commercially produced on enormous scales and resulted in widespread chemical pollution that can have detrimental effects on the environment. But a handful of artisans and textiles artists are making their own plant-based natural dyes at their kitchen tables, boiling up a dye bath on the hob using flowers or foods such as beetroot, berries, avocado skins, turmeric or saffron. Natural dyes result in very gentle hues and once the colour has been extracted, it must be set using a mordant or dye fixative, most commonly metal salts, so that any fabric doesn't lose colour quickly. The use of chrome salt as a mordant has now been banned, and aluminium salts are widely used in natural dyeing and deemed much less harmful. It seems somewhat ironic that the compound used to make a colour stick to the fabric can be more toxic than the dye itself.

When Australian textiles artist Ellie Beck started creating screen-printed fabrics by hand, she wanted to find sustainable alternatives to plastic tubs of paint. She began making small batches of fabric coloured with plant dyes, only collecting fallen and wind-blown plant parts or harvesting less than 10 per cent of the plant. 'Natural dyes are definitely slower to work with than synthetic dyes, and hues vary too, but that's the magic,' she told me. Beck is the founder of Petal Plum. To fix the dyes, she uses a mordant called alum (aluminium sulphate) or a plant's own tannins, but she also experiments with time as a mordant. She is aware, however, that mass production of natural dyes could strip the landscape of all colour. Her hope lies in green chemistry, which could allow access to rare natural colours without harming living things by replicating the vibrant pink sourced from cochineal bugs farmed commercially in South America or mimicking the distinct purple that is currently extracted by hand from sea snails.

Across the textiles industry, only 1 per cent of dyes used come from natural sources. So what are the pros and cons of synthetics versus natural dyes? Natural dyes simply don't produce bright colours that consumers love and they don't provide a colour range across all fibres. Finished products made with natural dyes tend to be much more subtle than any mass-produced coloured fabric made using synthetic dyes. Many people are sensitive or allergic in various degrees to different plants, and some foraged materials and garden plants are poisonous. If the production of natural dyes was scaled up, the harvesting of plant materials could contribute to deforestation and habitat loss. It's just not feasible in terms of land use and it would be resource intensive. For that reason, GOTS does allow the safe use of synthetic dyes and prohibits the use of natural dyes from endangered species.

Synthetic dyes can be produced relatively cheaply and easily in labs in the large volumes required by manufacturers, and can be specifically designed to be high performance. This means that the clothes produced with synthetic dyes are less likely to fade after a few washes. Of course, not all synthetics are bad. Durability and quality are key to not fuelling the appetite for more fast fashion.

Some kids wear Kapes

Millions of schoolchildren around the world are wearing fast fashion to school five days a week. School uniform is a multi-billion-dollar industry, but crease-free white shirts, firmly pleated tunics or trousers with hidden layers of Teflon-toughened fabric around the knees are not necessarily the best option for our children. Not only is the scale of production enormous, with huge environmental pollution, but children can be exposed to an alarming number of chemicals while wearing certain garments in the classroom. Tens of thousands of synthetic chemicals are used in the process of making a typical school shirt. Conventional cotton contains pesticide residues and materials get soaked in bleaches and dyes during production. Plus, there's the added issue of what happens to these uniforms once children have grown too big for them.

Fashion entrepreneur Matthew Benjamin founded Kapes in 2020 to establish a more sustainable and ethical supply chain for school uniforms that can be recycled. Based in Dubai, UAE, he produces school uniforms using GOTS-approved eco-friendly dyes, and uses materials such as organic cotton, Responsible Wool Standard-certified wool, recycled polyester (including ocean plastic) and regenerated nylon. Kapes recycles old uniforms to create new ones and ultimately develop a more circular economy.

'By switching from a polo shirt made from 100 per cent conventional cotton to one made using 100 per cent organic cotton, we could save half a kilometre in driving emissions, the equivalent of 68 light bulbs powered for an hour, and 191 litres of drinking water – per polo!' explained Benjamin, who is working with schools in UAE, Singapore, China, the UK, Thailand and Saudi Arabia.

Benjamin tested school uniforms being sold in the UAE and found that they contained some banned chemicals from a widely used group of synthetic colourants called azo dyes, which are most prevalent in black, brown, red, orange and yellow clothing. The chemical structure of these dyes hinges around benzene, which has been declared a 'major public health concern' by WHO. Long-term exposure to azo dyes, which make up 70 per cent of commercial dyes, has been associated with bladder and liver cancers. As early as 1895, workers in dye manufacturing were found to have an increased rate of bladder cancer. Azo dyes are regulated in the EU, USA and Canada, but not in many of the countries where production takes place, such as China and India. They're not restricted to textiles either – some azo dyes are used in cosmetics, food, tattoos and paper industries.

Some derivatives of azo dyes, which are commonly used to colour synthetic materials such as polyester, nylon and acetate, can be inhaled, leading to headaches and respiratory problems including asthma and rhinitis. Bacteria living on human skin can break down certain azo dyes to release cancer-causing compounds.

Children wear their clothes differently to adults – kids are much more active and their sweat could increase the rate of chemical absorption through the open pores in the skin, plus they tend to be more tactile, sucking and chewing on their clothes. Children can be more vulnerable to chemical exposure and wearing certain school uniforms made with synthetic dyes or other chemicals can trigger skin reactions. Reports show a rise in skin conditions such as atopic eczema and allergic contact dermatitis, and key chemicals responsible for this include epoxy resins which are used as gluing agents, flame retardants or UV-light absorbers. Formaldehyde is sometimes added to make the end product crease-free, Teflon is used to waterproof clothing and at least 980 endocrine-disrupting chemicals have been identified, such as BPA and phthalates, which are often found in waterproof clothing and flame-resistant fabrics.

In addition to exposure while wearing certain clothes, a 2015 report by the Danish Environmental Protection Agency revealed that children are exposed to volatile PFASs that evaporate into the air when wet waterproof gear is stored indoors and dried. While the researchers acknowledged that they were estimating worst-case scenarios, better ventilation could limit the impact of this 'off-gassing' from waterproof clothing, especially in school cloakrooms.[98]

Colour to dye for

The crux of the pollution problem lies in the fact that chemicals used throughout the textiles industry are being treated as single use. Phil Patterson, managing director at the UK-based Colour Connections textiles consultancy, explained that 'brands need to do a better job of policing pollution in their supply chains because most pollution from textiles comes from safe chemicals, rather than from toxic chemicals in textiles nowadays'.

TIPS FOR TOXIC-FREE TEXTILES

- **Make your clothes last:** learn to mend clothes instead of replacing them, and recycle fabric at your local clothes bank – don't send it to landfill.
- **Go vintage:** buy second-hand or do a clothes swap with friends – pre-loved clothing has fewer residual chemicals and reduces demand for the production of new items.
- **Shop less:** create a capsule wardrobe, buy clothes from brands that are meeting zero discharge commitments and choose natural fibres such as wool, hemp and organic cotton made with eco-friendly dyes.
- **Avoid added extras:** easy-iron, shrink-resistant, stain-resistant, crinkle-free, sweat-resistant and antibacterial elements all require extra chemical finishing agents.
- **Ask questions:** ask the shop assistant about a brand's sustainability credentials or email the manufacturer for information about the chemical processing used. Use Fashion Revolution's online template to ask a brand how an item has been processed before you buy it (fashionrevolution.org/about/get-involved).
- **Wash before you wear:** wash new clothes, bed linen and towels before first use. Fashion Revolution suggests soaking them overnight in bicarbonate of soda if you have time or adding white vinegar to your wash cycle to help remove formaldehyde.

When cotton is dyed, 70 per cent fixes to the fibres and approximately 30 per cent of the dye washes down the drain. Some newer dyes have been developed with up to 92 per cent fixation, which is much more efficient and saves water too, but that's an expensive option. Even when there is effluent treatment, that is often only being partially remediated and then solid waste still gets dumped into the environment.

In 2009, the Dutch start-up DyeCoo launched the world's first industrial dyeing machine that used carbon dioxide under high pressure to dye polyester, instead of water. The fabrics have to be washed before being dyed, so it's not entirely waterless. The dyes need to be made to incredibly high purities and it can't be used on a variety of fabrics, plus machines are prohibitively expensive, even for large textiles manufacturers, so this isn't something that could easily be scaled up. But a hybrid approach could be the solution. More synthetics could be made using plant-based ingredients rather than oil-based ones, and there are some exciting innovations in the textiles sector focused on reducing chemical pollution.

In Switzerland, a company called Archroma has developed a range of biosynthetic dyes called EarthColors made using non-edible agricultural waste such as leaves, residues from cotton plants or almond nutshells. These by-products are upcycled and transformed into commercially available dyes that can be used on fibres including cotton, linen, bamboo and viscose. The raw materials for these dyes are fully traceable along the supply chain and most are sourced from within 500 km of the production facility in Barcelona.

One dye that urgently needs a revolution is indigo. This prominent natural blue dye is traditionally made using the leaves of the indigo plant. One acre of cultivated indigo plants produces 5,000 kg of leaves and results in just 50 kg of pure natural indigo powder after processing. So for the enormous denim industry, synthetic indigo dyes are much more scalable and most blue jeans made today are dyed using synthetic indigo powder. But synthetic indigo also contains aniline, a toxic chemical that kills aquatic life, yet some 300 tonnes of it ends up in wastewater every year.

There are so many skeletons in the globe's fashion closet – from PFASs and phthalates to flame retardants and tanning chemicals.

In 2018, Archroma launched an aniline-free indigo dye and the accepted industry norm is starting to shift as aniline is beginning to appear on restricted substance lists. In Milan, an alternative 'smart indigo' system to process denim yarns has been developed by Luigi and Ilaria Caccia, the brother and sister co-founders of ItalDenim. Traditional dyes like indigo are usually insoluble in water and are made using polluting hydrosulphites. Smart indigo is created using only indigo pigment, caustic soda, water and electricity. It has a higher colouring power, so less dye is needed and it has a much lower carbon footprint because it's made on site and doesn't need to be transported. ItalDenim has developed a way to use chitosan to finish denim. This organic and biodegradable biopolymer is found in the shells of crustaceans such as crabs and lobsters, so it's a waste product of the food industry that can be used as a natural fixer that binds dyes, reducing the need for vast amounts of water and chemical dyes in the denim-making process.

The British company Colorifix uses a synthetic biology approach to make dyes. Bugs are bioengineered to reproduce the colours that are found naturally in living things. The process of fermentation using microbes replaces the need for toxic chemicals and reduces energy usage because dyeing can take place at 37° C rather than the conventional 150° C. The dye is effectively grown onto the textile. If this could be developed to match the performance of synthetic alternatives, perhaps it could be rolled out as a more energy-efficient option in the future.

In addition to the environmental pollution caused by chemical effluent from textiles factories, the fast-fashion industry is an enormous contributor to climate change. From the transportation and global distribution to water and energy usage, plus disposal of textiles waste once clothes are discarded, we need to push for greater transparency across the making of all textiles and reduce all environmental impacts. As consumers, we have to know whether a manufacturer is treating the effluent, reducing water and energy consumption, and managing the chemicals used to make a product.

Phil Patterson's landmark report on chemical circularity[99] highlights how sustainability within the fashion industry has, until now, primarily focused on the recycling and reuse of fabrics and fibres. Now, it's vital that the sector rethinks how chemicals are used to manufacture and recycle our clothes and textiles. The report estimates that at least 9.9 million tonnes of chemicals are used in the processing of 60 million tonnes of apparel and home textiles. Patterson argues that chemical use and its discharge must be reduced significantly and he proposes a new Chemical User Responsibility (CURE)

FIVE TOXIC-FREE FASHION LABELS

1. STANDARD 100 by OEKO-TEX: every component has been independently tested for harmful substances and is harmless for human health at point of sale. oeko-tex.com
2. STeP by OEKO-TEX: production conditions and factories are assessed for social responsibility, environmental criteria, optimum health and safety, plus chemical management. oeko-tex.com
3. Bluesign: the manufacturer adheres to certain strict criteria in sustainable processing. bluesign.com
4. Global Organic Textile Standard (GOTS): for textiles with at least 70 per cent organic fibres, usually cotton, GOTS outlines a specific list of chemicals that can and can't be used, water and energy consumption must be minimized and factories must treat any effluent. Member organizations include the Soil Association (UK), Organic Trade Association (US), International Association Natural Textile Industry (Germany) and the Japan Organic Cotton Association. global-standard.org
5. BEST: the German International Association Natural Textile Industry's BEST label bans the use of production chemicals such as azo dyes and only certifies textiles made with 100 per cent natural fibres; it has a certification scheme for natural leather made without chrome tanning, mainly in Europe. naturtextil.de

model, whereby each upstream supplier in the supply chain is responsible for the chemicals they use at the location in which they are used. So, all along the supply chain, every chemical ingredient should either be recycled or remediated and any discharge during wet processing should be dramatically reduced as well.

Of course, it's likely that a leak-free closed-loop system might never exist, but stringent measures can improve the environmental impacts of the textiles supply chains drastically. A variety of techniques will be needed to increase energy efficiency and reduce water usage and toxic chemical pollution. Patterson recommends that new 'chemical circularity' labels could be introduced to communicate these efforts to end consumers.

Making more conscious choices about the clothes that we buy – and how much we buy – can have far-reaching positive consequences and help to eliminate the use of so many chemicals in the fashion industry. As well as thinking about where clothes come from and how they are made, it's crucial that we consider what happens to them once we've finished wearing them. Some companies are designing 'circular' clothes that can be returned and recycled – it's now possible to lease a pair of jeans and swap them once they wear out – and rental platforms are on the up. This all reduces the volume of wastage, but a more environmentally friendly approach to chemical use is crucial to minimize pollution. Fast fashion urgently needs a makeover and comprehensive regulation is the best way to set a new, global trend.

TOXIC-FREE TAKEAWAYS

- More time is spent in our bedrooms than any other room of the house, albeit while sleeping, so the changes we make here will have a big impact on our exposures in the long term.
- Think about how fabrics are being used before you decide which to invest more in. Textiles such as bed linen and towels are right next to your skin; curtains aren't.
- Fast fashion fuels chemical pollution and that's all driven by over-consumption.
- While ingredients aren't listed for clothes, PFASs are often added to clothing as a 'bonus', so just watch out for gimmicks.
- Toxic chemicals run like an invisible thread through the fashion industry and using chemicals as single use, particularly in textiles, has to stop.

8

In the Garden

If you've got a shed, greenhouse, garage or an outside store, chances are there might be all sorts of chemicals lurking inside: tins of old paint, half-used varnish for your garden fence, motor oil or discarded neoprene wetsuits and other sports equipment.

Half-empty tins of paints, metal coatings, powder coatings and wood lacquers can all be a source of VOCs, so it's a good idea to store these outside the house, in a secure place out of reach of children. Anything that markets itself as being stain-resistant, water-repellent or weather-resistant will most likely contain PFASs too. In paint, PFASs are used as a binder and additive to make it flow and appear glossy, and in sealants and wood lacquers, PFASs give oil and water repellency. Pouring unwanted toxic liquids down the drain is never a good idea. Most local authorities offer a hazardous-waste-disposal service for things like old paint, weedkillers, solvents, bleaches and petrol. Some stuff won't be toxic right now, but may well have been manufactured using toxic chemicals, so by disposing of things correctly you can ensure that nothing leaches out into the environment. It's possible to limit the amount of chemicals in your DIY toolbox by reducing the number of different types of paints and adhesives you buy (opting for linseed oil as a wood preservative, for example) and in some cases taking a different approach altogether – simply fixing things together using screws and mechanical fasteners instead of using glues can dramatically cut down on VOCs.

DRIVING CHANGE

Every time a car brakes or turns a corner while driving on the road, the gradual erosion of synthetic rubber tyres releases microplastics into the air or the particles get washed down the drains into the waterways. With 1.4 billion vehicles in use globally, mostly in Asia, Europe and North America, tyre particles are the second most prevalent microplastic pollutant in the ocean.[100] A British start-up called The Tyre Collective is engineering a solution. When the prototype device is fixed to a vehicle in the lab, an electric charge captures 60 per cent of the tyre dust as it is produced, before it becomes pollution. Of course, to really minimize toxic pollution, use your car less, perhaps consider sharing an electric car or joining a car club, take public transport, walk or cycle. And while most cars still have combustion engines, remember to regularly pump up tyres to increase fuel efficiency and minimize toxic emissions when you do drive.

In the garage, you might find bottles of de-icer, motor oil and car antifreeze. The main active ingredient in car antifreeze is ethylene glycol, which tastes deceptively sweet but rather surprisingly leads to thousands of cases of poisoning every year from people drinking it. Like alcohol, it affects the nervous system and causes intoxication, seizures and possibly coma. Propylene glycol, also used as a synthetic food additive, is a safer alternative to ethylene glycol. Always clear up spills of antifreeze carefully, check your car for leaks, and store bottles out of reach of children and pets. Preferably all of these car maintenance supplies should be locked away safely.

How does your garden grow?

Most fertilizers consist of nitrogen, phosphorus and potassium (a combination of chemicals that's known as 'NPK') and the addition of these nutrients helps plants to grow. Without nitrogen in particular, growth can be stunted and plants won't thrive; they might turn yellow. Just as nutrient-rich fertilizers provide food for plants on land, so too do they encourage aquatic plants or algae to grow. So when rain washes these chemicals off the surface of farmland and into the waterways, a sudden increase in residues of nitrates and phosphates causes a process called eutrophication that can happen on a local scale in a tiny leat near your garden or on a vast scale across oceans. With more nutrient-rich minerals available, algae grow rapidly and form harmful algal blooms on the water's surface. These red tides prevent sunlight from reaching other water plants beneath the surface, so they can't photosynthesize. When these plants die, bacteria work hard to decompose this organic matter, using up oxygen in the process that other plants and fish need to survive. This 'hypoxia' creates dead zones of deoxygenated water which can kill aquatic insects, shellfish and fish, disrupting entire food chains.

Since the 1960s, more than 400 dead zones have been identified globally, including the Loire, Thames and Yangtze. In Norway in 2019, one harmful algal bloom killed 8 million salmon worth more than $70 million. The Mississippi River alone drains over 40 per cent of US land, bringing with it run-off from agriculture and industry. This has created an enormous dead zone that fluctuates in size across the Gulf of Mexico. In 2020, it measured 5,480 km^2 and decimated tourism and fishing industries.[101] While artificial intelligence could analyse satellite data and monitor harmful algal blooms, and seaweed farms could potentially extract pollutants and reduce deoxygenation, real solutions lie miles upstream where chemical run-off needs to be dramatically reduced.[102]

The use of human faeces as fertilizer contributes to dead zones and spreads contaminants on land too. Once treated by a sewage plant, human waste turns into a sludge called biosolids that's rich in nitrogen and phosphorus. Usually that either gets incinerated or landfilled. In the US, 47 per cent of biosolids is sold as fertilizer that's spread on farmland, gardens and lawns.[103] In 2021, the US-based grass-roots environmental group the Sierra Club found excessively high levels of various different PFAS chemicals in eight out of nine home-gardening fertilizer products that they tested in America, and linked this back to the use of pollutant-rich biosolids in the making of fertilizers.[104] Each fertilizer contained between fourteen and twenty PFAS compounds, despite some labels stating the contents were organic, eco or natural. PFASs have been found in fertilizers made without biosolids, and while the presence of PFASs doesn't alone indicate a direct health risk, it does show that, once in use,

FEELING HOT, HOT, HOT

Fire pits and bonfires are particularly inefficient as an outside heat source and can be terrible for the environment, depending on your choice of fuel. Instead of burning your garden waste, compost it or send it to your local recycling centre. Opt for commercially dried or 'seasoned' wood fuel that won't release smoke particulates in high concentrations like wet wood will do. To minimize environmental impact, buy logs from a sustainably managed local woodland or buy briquettes made from waste materials such as sawdust. Preferably avoid anything that's been imported. Biofuels such as bioethanol made from plants such as corn or sugar cane have lower carbon emissions, but their production involves heavy usage of land and water so they're not ideal either. Outdoor patio heaters release fewer particulates but are not energy-efficient, especially those that run on gas. Infrared electric heaters are better because they don't produce flames or involve fuel combustion and these tend to be more directional than gas heaters so less heat gets wasted. Short-wave infrared heaters are more energy-efficient than medium-wave ones. But ultimately, a thick blanket, extra jumpers and a warm drink are the least polluting ways to stay warm!

these forever chemicals soon become ubiquitous and once present can easily find ways into the food chain through the plants that we grow.

Alternatives to synthetic fertilizers are available, but they tend to be more expensive. Organic fertilizers are usually sold as soil conditioners and include trace elements such as zinc that plants need for healthy growth. They include seaweed-based products, horse muck, cow dung, peat-free compost and bonemeal. These are healthier for the environment and, if you're growing fruit and vegetables in your garden, will result in healthier home-grown produce. Another option is to change your planting and grow more native species of plants that will be better suited to the particular soil type and local climatic conditions, so they will require less chemical input. Legumes such as peas, clover and trellises of beans or lupines that are nitrogen-fixers – plants that naturally add nitrogen back into the soil through a partnership with good bacteria – help to enrich the soil as they grow too.

Poisonous pest control

What's actually in weedkiller? Glyphosate, the main active ingredient in Roundup weedkillers sold in garden centres around the world, is nasty stuff and there's simply no need to use it. The problem is systemic because glyphosate gets absorbed right into plants via the roots or leaves and can't simply be washed off with water before eating. So residues of some of the chemicals applied to our crops end up on our dinner plate too. That's not good news. Glyphosate is an endocrine disruptor and a probable carcinogen – it's widespread in our environment and has even been detected in women's breast milk. Risk of exposure is highest during its application to crops, lawns or gardens because it can be inhaled or ingested, and in three landmark lawsuits, US juries have awarded billions of dollars in damages to people who contracted cancer after using this chemical as a herbicide.

Over time, resistance develops in the weeds, diseases and animal pests that garden chemicals target, so it can be a struggle to ever get off the pesticide treadmill. Many farmers have to apply ever-stronger pesticides to continually ensure large enough yields. When it rains, excess chemicals get washed away into nearby streams, rivers and lakes, harming plants and animals that live in that water.

Pesticides are also toxic to people, and the potency of some active ingredients is shocking. Although obviously not designed for human consumption, paraquat is the most lethal weedkiller in the world. One sip can kill a person and there's no antidote. Though outlawed in more than fifty countries, this chemical is linked to thousands of farmer poisonings in low-to-middle-income countries. Often, this weedkiller is applied without the recommended protective clothing and stored in farmers' homes, sometimes in drinks bottles, within easy reach of children who might be oblivious to the dangers until it's too late.

While bans do reduce incidences of poisoning, getting rid of single pesticides is not necessarily the best long-term solution. If a pesticide is harmful to anything other than its target species in one country, it will be just as harmful elsewhere. All highly hazardous pesticides need to be avoided everywhere. Bans don't yet apply to exports and in the UK, where paraquat became prohibited in 2007, thousands of tonnes are still produced for export to the US, Australia, Japan and low-to-middle-income countries. A Public Eye and Greenpeace Unearthed investigation found that the UK is Europe's biggest exporter of toxic banned pesticides to poorer countries such as South Africa, Brazil and Ukraine, accounting for 40 per cent of the total exported.[105] Residues of illegal pesticides have been found in foods imported back to countries where these chemicals are not permitted. With global food-supply chains and international trade, there simply is no 'away', and it just doesn't make any sense to send the worst products to countries unable to afford alternatives. The baseline needs to be fair for everyone.

Even with the best intentions, your own home and garden can be affected by pesticides being applied by other people in your community. Pesticide drift happens when these chemicals are transported by the wind, like the droplets in an aerosol spray, and end up in places they weren't meant

PLUG-IN PLANTS

If you don't have a garden, you can still grow plants on a balcony or even on a windowsill if you don't have any outside space at all. You don't even need soil. Hydroponics is a hi-tech, super energy-efficient way of growing organic herbs, salads, fruits and flowers using a mineral-rich substrate that's made by dissolving essential nutrients such as nitrogen, potassium and phosphorus in water. Because the plant roots are provided with the exact things they need, no agricultural chemicals are required. With the addition of LED lighting on a timer, plants can be grown all year round in pod systems on your kitchen worktop or office desk. With hydroponics, plants tend to grow faster and produce a bigger yield than conventional growing, so you can grow fresh, seasonal cooking ingredients such as basil, chilli or rocket leaves without worrying about forgetting to water them.

to be – so when these chemicals are applied at scale from a plane or from large machinery, they spread far and wide, but equally when a neighbour sprays chemicals onto their lawn, some can waft across the fence and in through your open windows. Wherever you may live, if pesticides have been recently applied to grass in a playground or on a field you've just walked through, residues can come straight into your home on the soles of shoes. If you see pesticides being applied, perhaps talk to your neighbours or local landowners about it. In the US, some people use bright yellow pesticide flags to warn others that chemicals have just been sprayed, but that's probably not a sure-fire way to keep children and pets away. Some local laws stipulate that homeowners should give forty-eight-hour warnings to neighbours before using pesticides, so it's easier to avoid being outside when pesticide drift is most likely to occur.

If you do choose to use pesticides for a specific use, don't ever mix two products together, avoid using spray products on windy days and always make sure you don't breathe in these chemicals. Don't ever spray open flowers with pesticides because this could harm butterflies, bees and other pollinators. If only a few plants require treatment, buy a small bottle of ready-to-use formulation and apply it carefully to specific weeds with a brush. Store them in a locked cupboard out of reach of children, preferably in a garage or outbuilding. Best of all, choose alternative ways to control any weeds in your garden, allotment or vegetable patch. Some gardeners make a less toxic weedkiller using white wine vinegar mixed with a little washing-up liquid as the surfactant, but these home brews and other natural concoctions can harm a range of plants and animals, and, as with anything that has not been officially tested, it's hard to quantify how its use affects the environment. You might prefer to simply pull weeds out by hand.

A gardener's biggest bugbear

Animal pests such as slugs, aphids, greenflies, snails and moths can be even more of an annoyance than weeds. Things like slug pellets won't ever be a permanent solution and will be toxic for the birds, hedgehogs and other animals such as toads and frogs that feed on the slugs. Some gardeners use 'slug pub' traps with beer to collect unwanted visitors, barriers of wool pellets or place bramble stems around the edges of raised beds to protect vegetables from slugs and snails. Plants with hairy leaves tend to deter slugs but, once again, removal by hand is effective. The best time to find slugs is at night with a torch, but they do have a strong homing instinct, so release them a good distance from your garden. Remember, though, that they aren't all bad news; leopard slugs are territorial and do attack other slugs in the garden, plus many types of slugs and snails feed on rotting plant material so they can aid decomposition and play a key role in your garden ecosystem.

Crop protection netting will prevent butterflies from laying their eggs on your plants and it's a good idea to encourage insects' natural predators by putting up bird boxes and feeders. Use traps, barriers, covers and better fencing rather than chemical-laden pest control such as mothballs to deter rodents. Less toxic solutions include fumigant products based on garlic extract, which are sold to deter insect pests from greenhouses, or refined rapeseed oil, which is used to block the breathing pores of small insects like aphids and mites but won't harm bees or ladybirds. Encouraging garden wildlife, such as hedgehogs, birds, frogs and toads, helps to keep pests under control, and pollinating insects like hoverflies predate on pests, so grow more native nectar-rich plants that encourage them to your garden.

TIPS FOR PLANET-FRIENDLY PETS

- Look out for chemical additives and artificial colours in pet food, and if it contains palm oil, check that it's certified sustainable.
- If you buy canned pet food regularly, check it's bisphenol-free.
- Use flea treatments sparingly when recommended by a vet.
- Find more eco-friendly products that could be healthier for your pet. In the UK, The Solid Bar Company has a pet-care range that includes vegan dog shampoo and equine bug repellent, and Cooper & Gracie's anti-flea and anti-tick spray is plant-based.
- Choose dog-poo bags that are certified compostable or simply use old newspaper to pick it up and bin it.

Fipronil was found in 99 per cent of the water samples taken from English rivers, and imidacloprid was discovered in 66 per cent of samples, despite both chemicals being banned for use on farms in the UK. The washing of pets results in these chemicals being flushed into the sewers.[110] But the regular use of flea treatments isn't essential and this toxic topical application doesn't need to be a monthly habit; always check with your vet and see if an oral medicine is available. Avoid any pet shampoos that claim to guard against fleas and ticks, and instead of using tick collars that are doused in pesticide, ask your vet about getting your pet vaccinated against Lyme disease.

Design and build

If you move to a new property or decide to renovate, it's really useful to find out what your patch of land was previously used for. Was it built on non-organic farmland? Is it located near old factories? Try to piece together a picture of what possibly might be left in the soil today and plan your garden accordingly.

If you're moving into an older building dating back to the 1970s or previously that hasn't been renovated, check for lead paint on the inside and out. If you suspect it may be lead-based, use a home lead-testing kit or arrange a professional test to find out for sure. Over time, lead paint deteriorates and peels, creating toxic dust. Lead paint needs to be professionally removed before redecorating with water-based paints, preferably before you move in.

When it comes to building work, think outside the box. So many of the building materials used during construction contain chemicals that aren't healthy for us or the environment. Find ways to design a healthier extension, garden office or garage block. Ask your architect or builder about the content of the building blocks or the renders that they plan to use and find out whether the intended insulation is laced in flame retardants. They might not know but the manufacturer will, and the whole supply chain needs to become more transparent. Question them to find out about safer alternatives and remember that a well-insulated, energy-efficient home will be cheaper and less polluting to heat. Currently, 'better' options tend to be niche and more expensive, but with more robust building regulations it will become easier to reduce the unnecessary use of certain classes of chemicals and improve access to less toxic materials.

We need a floor-to-ceiling approach and the US-based Green Science Policy Institute is calling for manufacturers to take responsibility and reduce the unnecessary use of PFAS chemicals in everything from sealants to membrane materials. PFASs are used on asphalt roofing as a coating to reflect more sunlight, on metal roofing to guard against corrosion and colour loss, and on tensile roofs to make them more flexible. When it comes to solid building materials like gutters, roofing, weather-proofing membranes, even roofing nails, the toxicity really occurs during its production in the factory.

Eco-conscious choices will reduce your family's toxic load, and make your home and garden a healthier place.

PFAS chemicals are used to make many durable, glossy varnishes, weatherproof masonry paint and UV resistant, rainproof fence treatments that protect against decay. Sometimes the benefit will outweigh the risk when it comes to waterproofing, stain resistance, corrosion prevention or friction reduction. But some of these added extras aren't always essential. PFASs are used in paint as binders and to give a smooth finish, and enhance the oil and water repellency of wood lacquers and the grease resistance of sealants. They are incredibly useful and effective, but when exposed to the elements constantly, what's the cost to the environment?

PFASs are present in storm water – water that comes from rain and melted snow – but not enough is yet known about how much leaks out from building materials into wastewater and urban run-off. Some of the PFASs used to make building materials include big molecules such as fluoropolymers that are used as coatings on roofing and these don't tend to release PFAS when it rains. Smaller residual PFAS molecules are more likely to wash off substances that are used to add water repellency or used to make inks, lacquers, varnishes and paints. So when rain pours down the fence or along silky smooth painted walls, slowly but surely PFASs could be leaching out into our surroundings.

Once in the water system, it's expensive and challenging to filter out PFASs and other chemical contaminants, and even then only high temperature incineration would actually destroy the collected pollutants. If they are not removed, they will eventually end up in streams, rivers and sea. When the ocean waves crash, some PFASs get released as marine aerosols or droplets in sea spray. These sea spray aerosols are the biggest source of atmospheric PFAS chemicals.[111] In the Great Lakes area of the US, PFASs are showing up in rainfall at much higher concentrations than other contaminants such as heavy metals, pesticides and PCBs. Most urban sites have considerably greater quantities of PFASs than remote and rural areas, but the pathways these chemicals take aren't yet fully understood.

Renovations never last for ever, but some chemical pollution does. The design of sustainable building materials is a more feasible option long term. Otherwise, when a building is demolished, what happens to the chemicals in the materials? If incinerated, emissions could contain toxic chemicals. If landfilled, they might seep out over time. With greater awareness and increasing demand for toxic-free alternatives from consumers and people working across the industry, the norm is beginning to shift. Kingfisher, the international home improvement company with more than 1,300 stores across Europe, has pledged to phase out phthalates, PFASs and halogenated flame retardants from its supply chain by 2025.

The chemical ingredient fluorine is an indicator that a product contains PFASs, so a good rule of thumb is to avoid anything labelled as perfluoro, polyfluoro or fluoro which indicate they contain PFAS and, if the ingredients are not listed in full, ask the retailer or manufacturer. Roofing materials can be made with silicone- or acrylic-based coatings. Acrylics can be used to make paints durable and glossy while silicones or paraffin waxes can add water repellency to wood lacquers. Silicones and epoxy resins can seal porous materials on building exteriors instead of PFAS-based concrete sealers. Linseed oil can be used as a simple wood coating and instead of adhesives, perhaps opt for the humble screw if appropriate. Something to think about next time you renovate, perhaps?

From the very fabric of your home to everything in the garden around your property, we do have some level of autonomy about the toxic chemicals we install and use on an everyday basis. Take a holistic view and think about every person and pet living in your home or using your garden and how their exposure might vary. Consider how you use your outdoor space too. Do you need an immaculate, weed-free lawn for children to play football on? Will you be eating any of the plants grown in your garden? Take an integrated approach and make the most of what's naturally available, from maximizing on the rainfall to composting the vegetable scraps, and perhaps even embracing the nettlebed as a haven for butterflies. A few simple steps could not only encourage nature into your green space, but drastically reduce your chemical footprint.

TOXIC-FREE TAKEAWAYS

- Think of your outdoor space as a microcosm of the planet – just as excess chemicals leach off farmland, so too will they run off flower beds and vegetable patches into your local waterways.
- The healthiest garden is an organic one, so ditch the synthetic fertilizers and pesticides and, over time, resilience will improve and biodiversity will thrive.
- Pets are often considered part of the extended family – don't forget to consider their toxic load, from food to flea treatments.
- Many toxic chemicals are hidden among DIY products, from PFASs in waterproofing products to VOCs in paints and adhesives, so choose carefully and wear a mask during application.
- If you're moving to a new house, retrofitting or building an extension, take this opportunity to audit the toxic chemicals in the construction materials you use and shop around for safer alternatives.

PART THREE

A Better Future

9

Think Circular

From the moment chemical ingredients are mined, harvested or concocted in a lab, they have an impact. Often, products are shipped around the world before they reach us and once we flush them or throw them away, they don't magically disappear. With all the complexities that come with global supply chains, chemical cocktails and subsequent knock-on effects, it's critical that we reduce consumption of toxic chemicals in order to curb our environmental footprint on the planet.

More than 100,000 synthetic chemicals exist on today's market and, by volume, almost two-thirds of those are classified as hazardous to human health or the environment.[112] While pollution prevention and remediation are both vital, changing the chemicals we're using in the first place is the most promising tactic. But even if we transformed manufacturing worldwide today and stopped producing toxic chemicals altogether, we'd still have a big problem on our hands – historical waste and the huge backlog of electronics, plastics, materials and pollution that has built up since industrialization began in the eighteenth century. A healthy environment is fundamental to a thriving economy and a sustainable future, so we urgently need an innovative mix of science and creativity to develop solutions. That begins with a circle, because what goes around comes around. Every action has an impact and it's time to design out waste and pollution.

Closing the loop

Our current economy is based on a linear model of take-make-dispose. We take raw materials from the environment, we make them into different things and then we dispose of them as waste. That's just not an efficient use of finite raw materials and we can't go on like this for ever. A genuinely circular economy involves designing products that can be reused, repaired and recycled while eliminating waste. So at the end of their life, materials, including chemicals, should be recovered and reused. Ideally, toxic chemicals need to be replaced with safer, cleaner, affordable alternatives that can biodegrade in order to make reuse more feasible.

Rather than chemicals of concern leaching out of factories, systems need to keep production chemicals in use for a longer time. In an ideal world, a closed loop system will capture, treat and reuse any effluent. In the food sector, indoor vertical farms have closed systems that recirculate resources. They not only use much less water and energy but don't need agricultural chemicals. Because everything is controlled, monitored and contained, nothing escapes. But in most set-ups, even if leaks and spills are avoided, any chemical waste will eventually have to be dealt with. So it makes more sense to avoid hazardous chemicals in the first place.

As well as using chemicals for longer within processing and finding efficient ways to recycle them, we need non-toxic waste streams to make materials more recyclable in the future. Waste is a resource, but when materials are broken into pieces and re-formed into new items, the chemicals that were in the original product remain. When wastepaper gets re-pulped to produce recycled paper, chemicals such as BPA stay at consistent levels for decades.[113] As more materials start to be recycled and repurposed, materials will get used in different ways that go beyond their intended use. And each material contains different combinations of numerous chemicals.

Take plastic bottles that are produced using phthalates and bisphenols. These are relatively easy to recycle. But if this plastic were to be used to make baby toys, the toys would end up containing chemicals that are banned from baby products in many countries. So it's risky for a company to opt for recycled materials, especially while virgin materials are currently so much cheaper.

If toxic chemicals can't be collected for reuse or broken down into harmless constituents, they need to be landfilled or destroyed.

Nanos could also make recycling complicated, especially without accurate labelling, so they should be reserved for crucial medical innovations or to filter out heavy metals at wastewater treatment plants. Adding antimicrobial nanosilver to socks, everyday plasters, gym leggings or period pants isn't logical.

Uncertainty about a material's chemical content is a huge barrier to circularity. What goes in must come out, so by substituting hazardous chemicals with safer alternatives at the very start of the manufacturing process, the scope for circularity is much bigger. To make recycling easier in the future, we need to avoid chemicals of concern in the first place. Recyclers need to know exactly what they're dealing with and fluorescent markers or electronic tags could improve traceability in the future.

Not only can we change how we grow and make stuff, we can consume differently and make things last much longer. Less owning and more sharing results in having to produce less in the first place, and libraries of things and fashion rental services are popping up everywhere.

Cutting-edge chemistry

Green chemistry opens up possibilities for safer, less toxic alternatives in the design, manufacture and use of chemical products. Some solutions lie in unavoidable food waste. Inside the peel of an orange, a green chemist can find polymers, starches, cellulose, proteins, waxes, lipids, flavours, fragrances, antioxidants and vitamins.

Dr Avtar Matharu, who works at the University of York's Green Chemistry Centre of Excellence in the UK, sources the chemical ingredients he needs from the leftover fruit skins: 'These renewable by-products provide me with everything I need for my formulations. I can retune what nature gives us so I don't need to resort to crude oil.' According to Matharu, orange peel has so much potential: 'When you peel an orange you can smell it and that aroma comes from chemical compounds. Orange oil is a fantastic household cleaner, the pith contains pectin which people use for making jam and the peel stays orange for a long time thanks to a stable natural colourant. The inside of orange peel holds a large quantity of liquid, so it is actually a fantastic hydrogel too, just like the chemicals used in a nappy.'

Matharu's spin-out company has developed a material known as Starbon that can separate out compounds from fruit and vegetable waste, as well as purify water and trap gases, in the same way that activated carbon in odour-eaters and cooker hoods have tiny pores that absorb noxious gases and other chemicals. Named after the combination of 'starch' and 'carbon', it acts like a solid sponge. The pores can capture and release noxious gases and harmful chemicals.

Green chemistry is the design of chemical processes and products that reduce or avoid the use of hazardous substances and harmful end products. The term was first adopted by the EPA in America in the 1990s, primarily with a focus on pollution prevention. Rather than treating chemical waste after a material has been produced, green chemistry tries to prevent the production of waste in the first place, avoid hazardous chemical processes and create final ingredients that will easily degrade.

Paints have been made from soya oil and sugar, while antioxidant-rich coffee grounds can be turned into nutritious face scrubs. New packaging made from seaweed or unwanted fish waste is being designed to disappear completely. The dried blackcurrant skins left over after fruit has been pressed for Ribena are being used to make biodegradable purple-berry hair dyes and the scope for transforming the cosmetics industry using green chemistry is huge. Unlike the textiles industry, which aims to treat waste chemicals when clothes are manufactured before they're released into the environment, that's never the case with the cosmetics used in our bathrooms, hair salons or beauty parlours. Rather than reformulating existing products, green chemists begin by considering the end-of-life impact of every ingredient, and aim to make the finished product as environmentally friendly as possible.

Green chemistry could form the basis for a step-change in manufacturing, but feasible solutions not only have to be commercially and technically viable, they mustn't compromise on performance. So the necessary redesign is a big ask. The need for more sustainable ways of making, using and reusing things has never been greater, but rather than focusing on niche solutions or single steps in a complex, convoluted supply chain, true innovation is needed across entire industries and full life cycles. Rather than thinking of multiple, linear supply chains, these need to be integrated so waste from one process could be used as the raw material for another.

Chemical leasing is changing the single-use usage of chemicals such as solvents. Traditionally, manufacturers would earn more if they sold a chemical in greater volumes. Now, when chemicals are 'leased', suppliers get paid for the service that a chemical offers, so the amount earned won't depend on the volume of chemicals being used. Instead, the manufacturer gets paid for the performance of a chemical. This discourages overconsumption, saves companies money and reduces the amount of hazardous waste produced.

WHAT MAKES CHEMISTRY GREEN?

Green chemistry follows twelve key principles. Together, these result in:

- the prevention of waste and avoidance of spillages into the environment;
- the design of safer chemicals and products that biodegrade after use;
- the development of simpler, less hazardous chemical processes;
- the use of more renewable raw materials;
- an increase in overall energy-efficiency.

While environmental aspects are crucial to sustainability, sustainable chemistry needs to be economically successful and ethical for society too. The International Sustainable Chemistry Collaborative Centre (ISC_3) in Bonn, Germany, works with a Tanzanian start-up called EcoAct that transforms post-consumer plastic waste into insect-proof timber and building materials that won't rot. Meanwhile, in India, Banyan Nation is revolutionizing waste management by providing a livelihood to local waste pickers and using sustainable chemistry to clean up the contaminated plastics. By using an app that traces the materials through the supply chain, combined with technology that tests materials and uses them to make granules of plastic that rival virgin plastics, Banyan Nation is establishing a financially viable and ethical alternative to plastic production.

While collaborating with these start-ups, Dr Claudio Cinquemani, director of science and innovation at ISC_3 has noticed that entrepreneurs in low-to-middle-income countries often tend to consider waste as a resource. ISC_3 works with sisters, Jacqueline and Isemar Cruz, who started their venture, Le Qara, when they realized how much pollution resulted from the use of chromium in leather tanneries in their home town of Arequipa in Peru. They use microbes to transform abundant food waste into vegan leather that's biodegradable without using toxic chemicals.

'The transformation to more sustainable chemistry will happen, perhaps by 2030 – every business model that is unsustainable will have no future. I'm sure about that because at some point you will have to pay for carbon emissions, you'll have to pay for waste and you cannot rely on unethical work conditions any more,' Cinquemani told me. That begins with training the chemists of the future. ISC_3 has established an international school for sustainable chemistry, its first Master's degree in sustainable chemistry launched at Leuphana University in Lüneburg, Germany, and with 120 offices around the world, ISC_3 training programmes are underway across the Global South.

Green chemistry is already being embedded into the curriculum for some children. A teacher programme run by US not-for-profit Beyond Benign brings this subject to life in classrooms from early years right through to higher education. Not only will this help to create a workforce and scientific community that can expand on these green technological advances in the future, it equips all students to better understand the science behind the products they buy and become more eco-literate. Co-founder of Beyond Benign, Dr Amy Cannon holds the world's first PhD in green chemistry from the University of Massachusetts,

MAGIC MUSHROOMS

From oil spills to ocean plastic, some fungi and their mycelium or roots help to clean up chemical pollutants in what's known as mycoremediation. Fungi specialize in decomposition. By harnessing this power, fungi can be used to remove certain toxic contaminants from the environment, from heavy metals to pesticide residues, plastics and even antibiotics. Button mushrooms can remove cadmium and lead from polluted soils. A fungus called turkey tail can decompose organophosphate pesticides. In Mexico, scientist Rosa María Espinosa Valdemar has developed a way for oyster mushrooms to degrade disposable nappies.[114]

Boston. Cannon expects the entire chemical industry to shift towards greener chemistry. As such, building green chemistry education into science curriculums is a crucial step in creating future scientists, the next generation of eco-consumers and creative, critical thinkers who can further drive a systems change. A cleaner, more sustainable future hinges on that.

Green, sustainable chemistry opens up so many exciting possibilities for a circular economy and low-carbon future. By closing the loop, chemical waste can be eliminated to dramatically reduce pollution. By following the principles of green chemistry, the design, management and production of chemicals and the associated processes are becoming much more efficient. So many green solvents, clever biomaterials and high-performance ingredients are being developed. Yet, despite huge investment in making these safer, more environmentally friendly solutions, a quick glance at what's on the supermarket shelves shows that this frustratingly hasn't translated to the retail market at scale yet. It's still far from mainstream, so the real challenge is to make these the new norm. Let's hope that we'll see plenty more examples of circularity in every room of our homes one day.

10

Toxic-Free Principles to Live By

It's probably fairly apparent by this point that toxic chemicals are everywhere. Avoiding them completely is impossible, and some pose more of a threat to our health and the environment than others, but there are lots of positive steps we can take to minimize the impact. Perfection is not the goal, but transparency is. We need to be able to feel empowered by understanding what product labels say, and as well as trusting our governance and the safety legislations that are put in place. In a world of complex supply chains, full traceability is essential, so we must demand to know much more from companies about where and how something has been made. Only then can we make a fair judgement about how to vote with our wallets. It's crucial that at every step of the process claims are verified, quantifiable targets get measured and sustainability strategies evolve to continually allow for improvements. Our knowledge and understanding of chemical pollution is not static, and so much more remains to be researched. Now's the time to plan ahead.

There's no such thing as zero impact. Every single thing we do or buy or eat has an impact on the environment, on society and on our health. We don't live in isolation. But on the flip side, the answers lie in our interconnectedness. Our actions hold the potential to have huge, positive and powerful ripple effects. Whether we're a citizen, business owner or policymaker, we all have a role to play and a community to influence. Individual action is no substitute for policy change, but everyone can put pressure on retailers, manufacturers and leaders, and encourage regulators to take more action on chemical pollution to stop it at source. The system is broken, but it can be fixed. Solutions start way back upstream and prevention is always better than cure.

A manifesto for a toxic-free future

Let's expose the unexposed

So much pollution happens unintentionally. Blame isn't the answer. But enabling people to adopt safer, more sustainable practices is. Invisible pollution needs to be made much more visible so that everyone is aware of the dangers and the solutions.

Exercise your right to know

Risk assessments are based on science, not consumer opinion, and rightly so. But sometimes, formulations are carefully guarded as intellectual property and it can be very difficult to find out which chemicals are used to make a product. That's crucial if you know you're sensitive to certain ingredients or choose to avoid certain things for ethical reasons. We shouldn't need a degree in chemistry to be able to decipher the label, and much more clarity for consumers is required. Until the toxicity of ingredients such as nanos has been properly assessed, measured and regulated, we should be told whether they are in the things we buy. In the absence of data and stringent laws, manufacturers won't always err on the side of caution. While these decisions shouldn't have to come down to consumer choice, we're entitled to make our own judgement and choose accordingly.

Less is more

Our everyday chemical footprints can be drastically reduced by consuming less stuff. Buy less, use less, waste less. Just as that applies to consumers, so too should it apply to the chemical industry. Manufacturers should have to justify that every chemical ingredient and process is essential and that added extras really do offer a tangible environmental or health benefit. By leasing chemicals as a service and ditching the highly hazardous chemicals to ensure materials can be more easily recycled, supply chains can become much more efficient in terms of raw materials, energy and cost. That's an essential step in creating a more circular economy.

The polluter pays, not the public

The chemical manufacturers causing contamination need to take responsibility for remediation and preventative measures. Otherwise it just becomes someone else's problem. That relates as much to waste effluent discharged from the factory as it does to packaging that gets thrown away once a product has been used. Deposit return schemes are one step forward. This extended producer responsibility can be applied to everything from e-waste to plastic bottles and ensures that raw materials can be recovered. More stringent legal standards create an even playing field. Perhaps we need a chemicals tax?

Eliminate the hazards, not the risks

At the moment, so much management of chemicals like pesticides is still heavily focused on risk. Any risk depends on three things: the hazard a chemical poses to human health or the environment; the exposure a person, community or landscape experiences; and the vulnerability or chance of damage. But that doesn't take unintended use into account or reflect the fact that we are exposed to hundreds of different chemicals at any one time. Who hasn't ever heated up leftovers in a plastic container? How many times do children play with things that aren't officially toys? Is every cleaning product in your home always kept under lock and key? Highly hazardous chemicals really must be banned.

Re-evaluate and reinvent

Chemicals of concern that were registered and permitted decades ago need to be automatically reassessed at regular intervals as our understanding of toxicity develops. The entire life cycle of a substance must be taken into account. As it stands, there are just way too many missing pieces of the jigsaw puzzle. Those gaps have to be filled and chemical production urgently needs to evolve. True innovation starts at the design stage, so we must go back to the drawing board rather than switch to regrettable substitutes that might have equally toxic effects.

Look to REACH

The EU leads the way in forward-thinking chemical policy and takes a precautionary approach with REACH (Registration, Evaluation, Authorization and Restriction of Chemicals) by assuming that a chemical is not safe until proven otherwise. The EU's latest sustainability strategy indicates that regulators are starting to approach chemicals in terms of whole classes or categories and that interactions between chemicals are being considered. There isn't time to consider every single one of the many thousands of chemicals on an individual basis. Some of the worst offenders are beginning to be phased out; that includes PFASs and endocrine-disrupting chemicals in consumer products.

Then look beyond

Globally, we need more comprehensive regulation of marketing language and of the safety standards, with harm to both human health, wildlife and the natural environment all taken into account. Even though the EU regulations on chemicals are the most stringent in the world, that's not enough. Countries such as Sweden, Norway, Denmark and Germany are pushing for an EU-wide ban on all PFASs by 2030, but we need global treaties and enforcement of strict pollution limits to enable comprehensive change. While countries such as China and Korea have been strongly influenced by the EU in their development of national chemical regulation programmes, imagine what could be achieved if best practice was adopted around the world?

Look to the future

Legislation can take years. We need to take a long-term, holistic view and put preventative measures in place now to minimize our exposure, avoid the production of certain chemicals and promote safer alternatives. Remember that it's not just about the safety credentials of one ingredient – we have to consider dose, long-term exposure and possible accumulation plus interactions with other chemicals. Exposure to emerging pollutants must be stopped before it's too late. Remediation of contaminated legacy sites requires more investment and we need to push for chemical justice to protect the most vulnerable communities. Legacies live on.

Take advantage of tech

Digitalization can speed up the transition to a toxic-free future. Blockchain technology can be used to track conflict minerals and nanomaterials throughout a circular system. Artificial intelligence can help to predict the toxicity and degradability of chemicals, and enable green chemists to design safer alternatives.

Realize the bigger picture

While individual active ingredients might be safety-checked, formulations might not be. Chemicals are not used in isolation under lab conditions, but rather in real-life scenarios. Replicating the chemical cocktail that a person is exposed to throughout their lifetime is difficult to do. More comprehensive testing and long-term monitoring is needed to investigate the subtle effects that toxics can have within our bodies throughout our lifetime. We need an ambitious systems approach to assess chemical combinations and their cumulative impacts

Bans work ... sometimes

Accidental poisonings have decreased in countries that have banned highly hazardous pesticides such as paraquat. Blood levels of PBDEs (polybrominated diphenyl ethers) have dropped dramatically (by 39 per cent within a decade) when flame retardants are prohibited. But still some illegal pesticides end up on imported foods, and so loopholes need to be addressed

WHEN IS A HAZARD JUSTIFIED?

The most hazardous chemicals need to be prioritized for essential use only if there are no safer alternatives available. But how should 'essential use' be defined? It starts with the endgame. Chemicals that are used to make toys, luxury items, cosmetics and so on should be considered non-essential. We can survive without them. Furniture, cleaning products and clothes are essential, but the use of hazardous chemicals in them is not. So adding PFASs to board shorts to make them water-repellent is crazy considering they'll be used by surfers in the sea and are meant to get wet. EU REACH regulations aim to define criteria for 'essential use' and implement this in its chemical policies, which could prove to be a real breakthrough.

Scrub up on science

Follow the facts. Online commentary is often based on opinion. Doubts about the chemicals in our cosmetics loom far larger than worry about the banned hazardous pesticides we're eating on a daily basis or the PFASs we're exposed to – perhaps it's because the 'natural' alternatives for personal care products represent a growing market share and there's money to be made. So keep things in perspective. It's not fair that consumers are being manipulated by anecdotal evidence or emotional associations – we need to lift the lid on *all* of the toxic chemicals in our everyday lives, not just the ones that companies want us to know about. Let's educate our children and teach a new approach. Inspiring the next generation of innovators, green chemists and ethical entrepreneurs is a fundamental part of our transition to a low-carbon future.

Above all, we need to have foresight. When toxic chemicals like PCBs were first created, they were welcomed as innovation. Now, decades later, the full extent of forever chemicals is becoming apparent and some toxic chemicals are causing havoc on a global scale. While waiting for confirmation that most PFASs will persist in the environment indefinitely and accumulate in our blood, or that pesticide residues could act differently in combination, or that nanos are more impossible to remediate than microplastics, we can reimagine how we go about our normal daily lives and consider the responsibility we have to future generations.

Here's my ten-step plan to reducing toxic chemical pollution today:

1. **Join the dots.** Think about how something has been manufactured and how it will be disposed of. Where will it go once you've finished using it? Buy less stuff in the first place. Repair things when they break. Do your best to avoid food waste and make reusable, refillable products a habit instead of single-use items. Use smaller squirts and dollops of products wherever possible. Be resourceful in every way.

2. **Ask more questions.** Be more curious about labelling. Request more information from manufacturers, look for proof of any obscure claims, ask for full ingredients lists if they're not available. Hold companies to account publicly via social media, write letters or emails, or sign up to campaigns. Just don't be silent.

3. **Stay vigilant.** Don't get duped. Be aware of greenwash and avoid the common pitfalls. Some things are just clever marketing ploys. Download an eco-conscious shopping app (see page xx) so you can quickly decipher labels, verify claims and make better choices.

4. **Smell the roses.** Or the coffee or the fresh loaf of bread. Just go fragrance-free. Cut out those synthetic smells and ventilate your home every day.

5. **Reduce your toxic load.** Streamline the number of chemicals that get used in your home and use smaller amounts of each one.

6. **Consider price in terms of value.** If something is cheap, someone or something is paying the price

for that somewhere along the line. Invest more in things that will last and that have been designed with end-of-life in mind. If we buy less but buy better, we'll be more likely to waste less.

7. **Be climate-conscious.** Petrochemicals are one of the largest drivers of global oil demand, plus burning fossil fuels contributes to global heating and pollution. By buying less stuff, flying less, reducing food waste and being more energy-efficient at home, we can reduce pollution too.

8. **Be an agent for change.** Chat to your friends, colleagues, children, grandchildren, nieces and nephews about the changes you're making to reduce the negative effects of toxic chemicals, and lead the next generation of eco-conscious citizens.

9. **Disruption begins with us.** Remember that what's on the supermarket shelves can give a very distorted view of what's available and what's best. Just because it's being offered in multiple varieties does not mean it's ideal. And much of the available choice is all but an illusion – most mainstream brands link back to just a few large companies. Search for other options.

10. **Start small.** Make sustainable switches that work well for you. Start with just one change that you enjoy and make it a habit.

Life gets busy and everyone has a different set of priorities, so remember that perfection is impossible. Next time you need to replace something, think about whether its production, use and disposal might be toxic to the environment, wildlife and people, consider how essential it is to you and find out what the alternatives are, whether they are easy to use, as effective, more affordable and safer. Every decision will be a pay-off.

From having my blood tested to opening up the hoover and discovering what's in my house dust and then examining the readings on the air purifier, this has been a fascinating journey of discovery for me. With knowledge comes empowerment. The more we all know about the invisible pollutants in our homes, the more easily we can join those dots and change our ways for the better. Ultimately, that goes way beyond deciding which product to buy; it's about demanding a more streamlined, consistent, precautionary approach to using certain toxic chemicals in our everyday lives.

Your own toxic-free journey might start with streamlining one shelf, clearing out one cabinet or perhaps just rebooting one part of your daily routine. Fewer, more considered product choices can result in less waste and save you money too. By making a few hassle-free switches that are easy to sustain, positive change can be woven into your lifestyle and maybe one day we won't have to worry about toxic chemicals in our homes. Every room should be clean, safe and healthy. Collectively, we can make choices that are less toxic for people and the planet, now and in the future.

How to Find Out More

As quickly as more toxic chemicals come to market, new alternatives are being introduced. The landscape is constantly changing and more research is being published. To make sure you have the very latest facts at your fingertips, I have included a selection of useful resources comprising podcasts, campaigns, apps, books and films. So, if you're keen to delve further into this fascinating subject, here's a great place to start.

Chapter 2

Find out more about the health impacts of toxic chemicals:

- For the latest news about POPs, follow the UN's Safe Planet project on Facebook: facebook.com/safe.planet
- Discover more about labelling and chemical testing: chemicalsinourlife.echa.europa.eu
- The International Agency for Research on Cancer lists possible, probable and confirmed carcinogens: iarc.who.int
- Learn more about endocrine disruptors: chemtrust.org
- The EU's Freia Project campaigns for improved chemical safety testing that covers the complexities of female reproductive health to safeguard against endocrine-disrupting chemicals: freiaproject.eu

- Follow the progress of the US-based Taking Stock Study that investigates which products women use every day and how these contribute to disparities in health: takingstockstudy.org
- The Global Alliance on Health and Pollution publishes a map of lead pollution: lead.pollution.org
- Find medical professionals who specialize in environmentally acquired illnesses: iseai.org
- Take the QEESI test for toxicant-induced loss of tolerance: tiltresearch.org
- The Green Science Policy Institute's four-minute videos highlight classes of chemicals including PFASs, flame retardants and antimicrobials: sixclasses.org
- Listen to the podcast Talking PFAS: player.whooshkaa.com/shows/talkingpfas
- The Children's Environmental Health Network parent-educator toolkit includes storybooks about lead hazards and lesson plans for teachers: cehn.org
- HEAL, the Europe-based Health and Environment Alliance, campaigns for toxic-free economies and ways of life for more than 200 million people in fifty-three countries: env-health.org
- The Forum of International Respiratory Societies promotes respiratory health and highlights the need to reduce air pollution, including PM2.5, globally: firsnet.org

Chapter 3

Find out more about how chemical pollution affects the environment:

- Easy-to-read fact sheets about chemical pollution: who.int
- The International Pollutants Elimination Network is a coalition of environmental-health groups: ipen.org

- Environmental law charity ClientEarth holds power to account, tackling pollution in fifty countries: clientearth.org
- The international non-profit Pure Earth runs remediation projects in low-to-middle-income countries where human health is most at risk: pureearth.org
- If you find nurdles on the beach, get involved with International Pellet Watch: pelletwatch.org
- Check out whether beaches near you have been awarded the Blue Flag status: blueflag.global
- Surfers Against Sewage's app, Safer Seas & Rivers Service, gives UK water-users real-time alerts about sewage pollution: sas.org.uk
- This interactive map shows how one raindrop travels to the sea from any point in the US: river-runner.samlearner.com
- The HotSpots H2O series highlights water-crisis stories from around the world: circleofblue.org
- An EU-wide alliance working to combat human and animal antimicrobial resistance: saveourantibiotics.org
- The Pesticide Action Network consists of 600 organizations in ninety countries working to phase out hazardous pesticides: pan-international.org
- Public Eye, the Swiss NGO that investigates corporate wrongdoings, has published various in-depth investigations into pesticides: publiceye.ch
- Compare your air-pollution footprint to the national average with the personal air-pollution calculator created by UK charity Global Action Plan: calculator.cleanairhub.org.uk
- Plot air pollution where you live: stateofglobalair.org/data/#/air/plot
- Download the clean-air app to get an air-quality forecast for the next twenty-four hours: map.airly.eu
- A global campaign for cleaner air: breathelife2030.org

Chapter 4

Find out more about the products in your bathroom:

- The US-based Breast Cancer Prevention Partners' Campaign for Safe Cosmetics publishes a red list of chemicals of concern: safecosmetics.org
- Cosmetics Info is a comprehensive database of ingredients sponsored by the Personal Care Products Council: cosmeticsinfo.org
- The Protect Land + Sea certification independently tests sunscreen brands for toxic chemicals: haereticus-lab.org/protect-land-sea-certification
- Campaign group Women's Voices for the Earth has created a timeline of chemical testing in period products since 2002: womensvoices.org
- The UK's Cosmetic, Toiletry and Perfumery Association publishes a useful directory of ingredients: thefactsabout.co.uk
- Use this US Environmental Protection Agency search facility to find products that meet the Safer Choice Standard: epa.gov/saferchoice/products
- MADE SAFE has published a hazard list to highlight the worst chemical offenders from A to Z: madesafe.org
- The UK's Society of Cosmetic Scientists runs Scrub Up On Science courses for schoolchildren: scs.org.uk
- Learn how to formulate your own skincare: schoolofnaturalskincare.com and formulabotanica.com
- Decode the ingredients listed on personal care product labels – incidecoder.com – or download the INCI Beauty app.
- Creators of the environmenstrual campaign, the UK's Women's Environmental Network has carried out a lot of research into health harms associated with period products: wen.org.uk

Chapter 5

Find out more about the products in your kitchen:

- The Consumer Product Information Database flags up chemicals of concerns listed on labels of 23,000 brands of consumer household products sold in the US: whatsinproducts.com
- Challenge major retailers to eliminate toxic chemicals such as PFASs from products and packaging: saferchemicals.org/mind-the-store/
- The Clean Label Project runs heavy-metal testing and a pesticide-free certification scheme in the US: cleanlabelproject.org
- Find PFAS-free alternatives: pfasfree.org.uk/current-initiatives/pfas-free-products or pfascentral.org/pfas-free-products
- The organization Women's Voices for the Earth has published a useful guide about how to quit the quats: womensvoices.org/cleaning-with-pesticides-quit-the-quats/
- For alternative cleaning products, take a look at these DIY recipes: womensvoices.org/take-actionwith-womens-voices/green-cleaning-parties/green-cleaning-recipes/
- Support charities that aid the transition to cleaner cooking stoves in low-to-middle-income countries: cleancookingalliance.org and solarcookers.org

Chapter 6

Find out more about the products in your living room:

- To find environmentally preferable office furniture, take a look at LEVEL certified products: levelcertified.org
- US-based campaign to protect children from air pollution and toxic chemicals: momscleanairforce.org

- Search this platform for consumer electronics and packaging made with safer alternatives: chemforward.org
- GreenScreen Certified furniture and fabrics are made with safer chemicals: greenscreenchemicals.org
- For in-depth reports and product ratings, Ethical Consumer magazine investigates everything from tech to textiles and is available internationally: ethicalconsumer.org

Chapter 7

Find out more about the products in your bedroom:

- Find a repair hub to mend your clothes at Mend Assembly's global directory: mendassembly.com/directory
- The Changing Markets Foundation sets out a road map for less polluting, more circular viscose production: dirtyfashion.info
- Every April, Fashion Revolution Week amplifies unheard voices from across global supply chains and calls for change: fashionrevolution.org
- The Good On You app uses expert analysis to give fashion brands a rating that reflects their sustainability credentials: goodonyou.eco
- Greenpeace's international investigation into toxic chemicals used in the fashion industry is worth a read: greenpeace.org/international/act/detox/
- Find out which outdoor clothing brands are ditching PFASs and which aren't: detox-outdoor.org
- The ZDHC is a collaboration of organizations from the textiles, footwear, leather and apparel industries involved in developing solutions to chemical management: roadmaptozero.com

Chapter 8

Find out more about the products in your garage and garden:

- Lots of great tips and safer alternatives to reduce pesticide use are published on this US campaign site: beyondpesticides.org
- World-leading soil consultant Jennifer Dungait works with farms to maximize soil health: soilhealthexpert.com
- IFOAM Organics International publishes up-to-date news about growing organic: ifoam.bio/
- In collaboration with Garden Organic, the Pesticide Action Network UK has published a free online guide to gardening without pesticides: pan-uk.org/site/wp-content/uploads/A-guide-to-gardening-without-pesticides-2017-bw-printer-friendly.pdf
- The Soil Health Academy runs training courses about improving soil health through regenerative agriculture: soilhealthacademy.org
- Search for a soil health consultant or train with Dr Elaine Ingham's Soil Food Web School: soilfoodweb.com
- The WELL Standard is a health-focused building rating system run by the International WELL Building Institute that aims to improve human health through design: wellcertified.com
- The Healthy Building Network uses a red-to-green ranking system to compare materials based on their hazardous content and recommends healthier alternatives: homefree.healthybuilding.net/products

Chapter 9

Find out more about how innovations in chemistry can reduce pollution:

- The International Chemical Secretariat publishes a list of about a thousand hazardous chemicals and encourages switching to safer alternatives: sinlist.chemsec.org
- In the EU, consumers have the legal 'right to know' whether a product contains any of the most hazardous chemicals of concern. The free Scan4Chem app can send automated requests that a company needs to respond to within forty-five days: askreach.eu
- Find out more about how to contact companies to check claims have been verified: askforevidence.org/help/guide-to-asking
- The globally recognized Cradle to Cradle Certified product standard ensures that the chemicals used are as safe for people and the environment as possible, and that they get reused in the manufacturing process: c2ccertified.org
- The Nordic Swan Ecolabel promotes a circular economy, from the sourcing of raw materials to disposal and recycling, without the use of certain hazardous chemicals: nordic-ecolabel.org
- The International Sustainable Chemistry Collaborative Centre offers various education training programmes, from Master's degrees to summer schools: isc3.org
- The American Chemical Society runs the Green Chemistry Institute: acs.org/greenchemistry
- Beyond Benign trains educators to integrate green chemistry into school curriculums: beyondbenign.org

Chapter 10

WATCH: Top films about toxics

Flint (2020) – A documentary about the true story of toxic drinking water in America, with Mark Ruffalo and Alec Baldwin.

Dark Waters (2019) – Featuring Mark Ruffalo again, this time the truth is revealed about Teflon.

Deepwater Horizon (2016) – A disaster film based on the Gulf of Mexico oil spill.

Stink! (2015) – A thought-provoking documentary about the chemicals in our consumer products.

Erin Brockovich (2000) – Julia Roberts plays the woman who uncovers a corporate scandal in her hometown.

READ: Brilliant books about toxics

Toxic Legacy: How the Weedkiller Glyphosate is Destroying Our Health and the Environment by Stephanie Seneff (2021)

Breathless: Why Air Pollution Matters and How It Affects You by Chris Woodford (2021)

Superman's Not Coming: Our National Water Crisis and What WE THE PEOPLE Can Do About It by Erin Brockovich (2020)

Clean Green: Tips and Recipes For a Naturally Clean, More Sustainable Home by Jen Chillingsworth (2020)

Children & Environmental Toxins: What Everyone Needs to Know by Philip J. Landrigan and Mary M. Landrigan (2018)

The Brown Agenda: My Mission to Clean Up the World's Most Life-Threatening Pollution by Richard Fuller (2015)

Acknowledgements

To all the scientists, entrepreneurs and campaigners around the world to whom I have spoken while researching this book – the biggest of thank yous. It is always such a privilege to speak to experts at the forefront of exciting new developments. I'm so grateful for your time, patience and guidance, because without that this book would simply be my interpretations and opinions. Now, it's a comprehensive guide to the very best advice I could uncover, the newest thinking and an explanation of how every part of that relates to us, in our homes, right now.

To my brilliant editor Jo Stansall, who had every confidence in me from the beginning, and the rest of the creative team at Michael O'Mara. To the Society of Authors and the Authors' Foundation for presenting me with the Roger Deakin Award for environmental writing. To my agent Amanda Preston at LBA Books for your support and encouragement. I appreciate all the wise words and professional insight as this book has evolved from a concept into a hardback.

To my wonderful parents, siblings and the entire Fox-Turns tribe. Thank you for cheerleading me along the way and having faith in my ability to translate some fairly hefty, geeky facts into a book full of stories and solutions.

To my very best friends. For helping me no end as deadlines loomed, for sending espresso martinis by post, for battle ropes and Paris zooms, sea swims and pep talks, cups of tea and hugs when I needed them most.

To Chris. Without you, this book simply would not exist. Your unwavering belief that I can achieve anything I put my mind to and your constant, calm support mean the world to me.

To Ella and Stanley. This book poses as many questions as answers. May you both always continue to ask 'why?' Know that your voices make an enormous difference – as children, as students, as consumers, as citizens and one day as voters. Own that agency and hold others to account whenever possible. Be the curious customer who questions the claims on a label and you'll see through the greenwash in all that you do.

Endnotes

1 Minderoo Foundation, 'The plastic waste makers index: revealing the source of the single-use plastic crisis' (2021). minderoo.org/plastic-waste-makers-index/findings/executive-summary/

2 World Wide Fund for Nature. tinyurl.com/3edj8hjw

3 European Environment Agency, 'Plastics, the circular economy and Europe's environment' (2021). eea.europa.eu/publications/plastics-the-circular-economy-and

4 International Energy Agency, 'The future of petrochemicals' (2018). iea.org/reports/the-future-of-petrochemicals

5 International Chemical Secretariat, 'What goes around' (2020). tinyurl.com/22td74eh

6 T. Legg, J. Hatchard, A. B. Gilmore. 'The science for profit model – how and why corporations influence science and the use of science in policy and practice', *PLoS ONE* (2021). tinyurl.com/2bmn55mt

7 Robin E. Dodson *et al.*, 'Personal care product use among diverse women in California: taking stock study', *Journal of Exposure Science & Environmental Epidemiology* (2021). tinyurl.com/z9k6djwt

8 European Environmental Bureau, 'Major safety gaps for most chemical files – official' (2019). tinyurl.com/mu4x52je

9 Philip J. Landrigan *et al.*, 'The Lancet Commission on pollution and health', *The Lancet* (2017). thelancet.com/commissions/pollution-and-health

10 M. Bellanger, B. Demeneix, P. Grandjean *et al.*, 'Neurobehavioral deficits, diseases, and associated costs of exposure to endocrine-disrupting chemicals in the European Union', *The Journal of Clinical Endocrinology & Metabolism* (2015). 10.1210/jc.2014-4323

11 Carl-Gustaf Bornehag *et al.*, 'Prenatal phthalate exposures and anogenital distance in Swedish boys', *Environmental Health Perspectives* (2015). tinyurl.com/3ujua3a5

12 *Chemical & Engineering News*, 'Research spending continues on an upward trajectory', 9 June 2019. tinyurl.com/ppdprshs

13 UNICEF, 'The toxic truth' (2020). tinyurl.com/k9knx7ua

14 J. Grigg, 'Environmental toxins; their impact on children's health', *Archives of Disease in Childhood*. tinyurl.com/yd547r7m

15 Helene Wiesenger *et al.*, 'Deep dive into plastic monomers, additives and processing aids', *Environmental Science & Technology* (2021). 10.1021/acs.est.1c00976

16 UK government statistics on emissions of air pollutants. tinyurl.com/2225h6kr

17 Ajith Dias Samarajeewa, Jessica R. Velicogna, Dina M. Schwertfeger *et al.*, 'Effect of silver nanoparticle contaminated biosolids on the soil microbial community', *Nano Impact* (2019). doi.org/10.1016/j.impact.2019.100157

18 S. B. Fournier, J. N. D'Errico, D. S. Adler *et al.*, 'Nanopolystyrene translocation and fetal deposition after acute lung exposure during late-stage pregnancy', *Particle and Fibre Toxicology* (2020). tinyurl.com/ejmj9yp7

19 European Commission, 'Plastics in the ocean'. tinyurl.com/8bcr893z

20 B. D. Laird, A. B. Goncharov and H. M. Chan, 'Body burden of metals and persistent organic pollutants among Inuit in the Canadian Arctic', *Environment International* (2013). 10.1016/j.envint.2013.05.010

21 Piera M. Cirillo, Michele A. La Merrill, Nickilou Y. Krigbaum and Barbara A. Cohn, 'Grandmaternal perinatal

serum DDT in relation to granddaughter early menarche and adult obesity: three generations in the child health and development studies cohort', *Cancer Epidemiology, Biomarkers and Prevention* (2021). 10.1158/1055-9965.EPI-20-1456

22 5 Gyres Institute, 'Plastic microbeads'. 5gyres.org/microbeads

23 Yukie Mato, Tomohiko Isobe, Hideshige Takada *et al.*, 'Plastic resin pellets as a transport medium for toxic chemicals in the marine environment', *Environmental Science & Technology* (2001). 10.1021/es0010498

24 C. Rochman, E. Hoh, T. Kurobe *et al.*, 'Ingested plastic transfers hazardous chemicals to fish and induces hepatic stress', *Scientific Reports* (2013). doi.org/10.1038/srep03263

25 The *Guardian*, 'Water firms discharged raw sewage into England's rivers 200,000 times in 2019', 1 July 2020. tinyurl.com/8s4288uj

26 Anne F. C. Leonard, Lihong Zhang, Andrew J. Balfour *et al.*, 'Exposure to and colonization by antibiotic-resistant E. coli in UK coastal water users: environmental surveillance, exposure assessment, and epidemiological study' (Beach Bum Survey), *Environment International* (2018). doi.org/10.1016/j.envint.2017.11.003

27 M. Lechenet, F. Dessaint, G. Py *et al.*, 'Reducing pesticide use while preserving crop productivity and profitability on arable farms', *Nature Plants* (2017). doi.org/10.1038/nplants.2017.8

28 United Nations General Assembly, 'Report of the special rapporteur on the right to food' (2017). tinyurl.com/tue7te7c

29 World Health Organization, 'Dioxins and their effects on human health' (2016). tinyurl.com/59ud4x6m

30 US Environmental Protection Agency, 'Volatile organic compounds' impact on indoor air quality'. tinyurl.com/4vyc3sut

31 N. Mueller *et al.*, 'Changing the urban design of cities for health: the superblock model', *Environment International* (2020). 10.1016/j.envint.2019.105132

32 M. M. Coggon *et al.*, 'Volatile chemical product emissions enhance ozone and modulate urban chemistry', PNAS (2021). tinyurl.com/yes4hwee

33 Qiang Zhang, Xujia Jiang, Dan Tong *et al.*, 'Transboundary health impacts of transported global air pollution and international trade', *Nature* (2017). doi.org/10.1038/nature21712

34 Global Alliance on Health and Pollution, 'Pollution and health metrics' (2019). gahp.net/pollution-and-health-metrics/

35 Statista, 'Value of the cosmetics market worldwide from 2018 to 2025' (2020). tinyurl.com/6pcuw78m

36 Campaign for Safe Cosmetics, 'International Laws'. tinyurl.com/yt74jfje

37 US Food and Drug Administration, 'Prohibited & restricted ingredients in cosmetics'. tinyurl.com/2catv8sm

38 CodeCheck, 'The invisible danger: Hidden polymers in cosmetic products' (2020). tinyurl.com/42jhu5fz

39 H. Whitehead *et al.*, 'Fluorinated compounds in North American cosmetics', *Environmental Science & Technology Letters* (2021). doi.org/10.1021/acs.estlett.1c00240

40 Qian Wang, Shumei Cui, Li Zhou *et al.*, 'Effect of cosmetic chemical preservatives on resident flora isolated from healthy facial skin', *Journal of Cosmetic Dermatology* (2019). 10.1111/jocd.12822

41 A. Aljarrah, N. G. Coldham, P. D. Darbre *et al.*, 'Concentrations of parabens in human breast tumours', *Journal of Applied Toxicology* (2004). doi.org/10.1002/jat.958

42 L. Barr, G. Metaxas, C. A. J. Harbach, L. A. Savoy and P. D. Darbre, 'Measurement of paraben concentrations in human breast tissue at serial locations across the breast from axilla to sternum', *Journal of Applied Toxicology* (2012). doi.org/10.1002/jat.1786

43 R. Hirose *et al.*, 'Situations leading to reduced effectiveness of current hand hygiene against infectious mucus from influenza virus-infected patients', *mSphere* (2019). doi.org/10.1128/msphere.00474-19

44 J. Gosse and L. Weatherly, 'Triclosan exposure, transformation, and human health effects', *Journal of Toxicology and Environmental Health*, Part B (2017). 10.1080/10937404.2017.1399306

45 Shaofang Cai, Chunhong Fan, Yingjun Li *et al.*, 'Association between urinary triclosan with bone mass density and osteoporosis in US adult women, 2005–2010', *The Journal of Clinical Endocrinology & Metabolism* (2019). tinyurl.com/xt55z8

46 Unilever, 'Triclosan and triclocarban'. tinyurl.com/2uxu5bns

47 B. Vande Vannet, B. De Wever *et al.*, 'The evaluation of sodium lauryl sulphate in toothpaste on toxicity on human gingiva and mucosa: a 3D *in vitro* model', *Dentistry* (2015). tinyurl.com/ndkspbyn

48 K. E. Boronow, J. G. Brody, L. A. Schaider *et al.*, 'Serum concentrations of PFASs and exposure-related behaviors in African American and non-Hispanic white women', *Journal of Exposure Science & Environmental Epidemiology* (2019). doi.org/10.1038/s41370-018-0109-y

49 Yousuf H. Mohammed *et al.*, 'Support for the safe use of zinc oxide nanoparticle sunscreens: lack of skin penetration or cellular toxicity after repeated application in volunteers', *Journal of Investigative Dermatology* (2018). doi.org/10.1016/j.jid.2018.08.024

50 C. A. Downs, E. Kramarsky-Winter, R. Segal *et al.*, 'Toxicopathological effects of the sunscreen UV filter, oxybenzone (benzophenone-3), on coral planulae and cultured primary cells and its environmental contamination in Hawaii and the US Virgin Islands', *Archives of Environmental Contamination and Toxicology* (2016). 10.1007/s00244-015-0227-7

51 *The New York Times*, 'Is your sunscreen poisoning the ocean?', 19 August 2017. tinyurl.com/t7mcnfjc

52 Statista, 'The US leads the world in toilet paper consumption' (2018). tinyurl.com/tbmjbkbs

53 ANSES, 'ANSES recommends improving baby diaper safety', 23 January 2019. tinyurl.com/3atvnvfn
54 Campaign for Safe Cosmetics, 'Cumulative exposure and feminine care products' (2016). tinyurl.com/3ws79rjf
55 European Parliament News, 'Plastic in the ocean: the facts, effects and new EU rules'. tinyurl.com/yuh3ryfw
56 Women's Environment Network, 'Seeing red: menstruation and the environment'. tinyurl.com/ayupad72
57 Women's Voices for the Earth, 'Always pads testing results'. tinyurl.com/yfk5fccm
58 Yao He, Yele Sun, Qingqing Wang *et al.*, 'A black carbon-tracer method for estimating cooking organic aerosol from aerosol mass spectrometer measurements', *Geophysical Research Letters* (2019). doi.org/10.1029/2019GL084092
59 S. Patel *et al.*, 'Indoor particulate matter during HOMEChem: concentrations, size distributions, and exposures', *Environmental Science & Technology* (2020). 10.1021/acs.est.0c00740
60 *South China Morning Post*, 'Nearly 100 million Chinese people supplied drinking water with "unsafe" levels of toxic chemicals', 16 January 2021. tinyurl.com/ptc4xvhd
61 Hannah Gardener, Jaclyn Bowen and Sean P. Callan, 'Lead and cadmium contamination in a large sample of United States infant formulas and baby foods', *Science of The Total Environment* (2019). doi.org/10.1016/j.scitotenv.2018.09.026
62 John Fagan, Larry Bohlen, Sharyle Patton and Kendra Klein, 'Organic diet intervention significantly reduces urinary glyphosate levels in US children and adults', *Environmental Research* (2020), doi.org/10.1016/j.envres.2020.109898
63 Pesticide Action Network, 'Pesticides in our food', pan-uk.org/dirty-dozen
64 Environmental Working Group, 'EWG's 2021 shopper's guide to pesticides in produce'. tinyurl.com/3kdz7fms
65 Pesticide Action Network, 'The cocktail effect'. tinyurl.com/3demcbbv

66 Pesticide Action Network, 'Banned and hazardous pesticides in European food: report highlights' (2020). tinyurl.com/2z4v5y58

67 Jenna Forsyth *et al.*, 'Sources of blood lead exposure in rural Bangladesh', *Environmental Science & Technology* (2019). 10.1021/acs.est.9b00744

68 Toxic-Free Future, 'Take out toxics'. tinyurl.com/33bk29sy

69 C. Janson *et al.*, 'The European Community Respiratory Health Survey: what are the main results so far?' *European Respiratory Journal* (2001). tinyurl.com/4muu3bcb

70 Elisabetta Caselli and Ivana Purificato, 'Could we fight healthcare-associated infections and antimicrobial resistance with probiotic-based sanitation? Commentary', *Annali dell'Istituto Superiore di Sanità* (2020). 10.4415/ANN_20_03_03

71 Mathew Jackson *et al.*, 'Comprehensive review of several surfactants in marine environments: fate and ecotoxicity', *Environmental Toxicology and Chemistry* (2015). doi.org/10.1002/etc.3297

72 Amanda L. Valdez, Marcel J. Casavant *et al.*, 'Pediatric exposure to laundry detergent pods', *Pediatrics* (2014). 10.1542/peds.2014-0057

73 Imogen E. Napper *et al.*, 'The efficiency of devices intended to reduce microfibre release during clothes washing', *Science of The Total Environment* (2020). 10.1016/j.scitotenv.2020.140412

74 C. A. Paterson, R. A. Sharpe, T. Taylor and K. Morrissey, 'Indoor PM2.5, VOCs and asthma outcomes: a systematic review in adults and their home environments', *Environmental Research* (2021). doi.org/10.1016/j.envres.2021.111631

75 US Environmental Protection Agency, 'Volatile organic compounds' impact on indoor air quality'. tinyurl.com/4vyc3sut

76 Brian C. McDonald *et al.*, 'Volatile chemical products emerging as largest petrochemical source of urban organic emissions', *Science* (2018). 10.1126/science.aaq0524

77 Anne Steinemann, 'Ten questions concerning air fresheners and indoor built environments', *Building and Environment* (2017). doi.org/10.1016/j.buildenv.2016.11.009
78 R. Chakraborty, J. Heydo, M. Mayfield and L. Mihaylova, 'Indoor air pollution from residential stoves: examining the flooding of particulate matter into homes during real-world use', *Atmosphere* (2020). doi.org/10.3390/atmos11121326
79 Junjie Zhang, Lei Wang and Kurunthachalam Kannan, 'Microplastics in house dust from 12 countries and associated human exposure', *Environment International* (2020). doi.org/10.1016/j.envint.2019.105314
80 Anna Kärrman and Ulrika Eriksson, 'World-wide indoor exposure to polyfluoroalkyl phosphate esters (PAPs) and other PFASs in household dust', *Environmental Science & Technology* (2015). doi.org/10.1021/acs.est.5b00679
81 Prabjit Barn *et al.*, 'The effect of portable HEPA filter air cleaners on indoor PM2.5 concentrations and second hand tobacco smoke exposure among pregnant women in Ulaanbaatar, Mongolia: the UGAAR randomized controlled trial', *Science of The Total Environment* (2018). 10.1016/j.scitotenv.2017.09.291
82 Steve McNeil, 'The removal of indoor air contaminants by wool carpet', *AgResearch* (2015). tinyurl.com/f7se8mx9
83 B. C. Wolverton and John D. Wolverton, 'Plants and soil microorganisms: removal of formaldehyde, xylene, and ammonia from the indoor environment', *Journal of the Mississippi Academy of Sciences* (1993). tinyurl.com/jy7p2axk
84 A. S. Young *et al.*, 'Impact of "healthier" materials interventions on dust concentrations of per- and polyfluoroalkyl substances, polybrominated diphenyl ethers, and organophosphate esters', *Environment International* (2021). tinyurl.com/36mz2zeh
85 UN Environment Programme, 'UN report: time to seize opportunity, tackle challenge of e-waste', 24 January 2019. tinyurl.com/4p8vsjv9
86 Arlene Blum and Bruce N. Ames, 'Flame-retardant

additives as possible cancer hazards', *Science* (1977). 10.1126/science.831254

87 SGS, 'Aiming for zero', 11 July 2017. sgs.com/en/news/2017/07/aiming-for-zero

88 Minakshi Jain, 'Ecological approach to reduce carbon footprint of textile industry', *International Journal of Applied Home Science* (2017). tinyurl.com/wdhxxv4e

89 Textile Exchange, 'Preferred Fiber & Materials: Market Report 2020'. https://textileexchange.org/wp-content/uploads/2020/06/Textile-Exchange_Preferred-Fiber-Material-Market-Report_2020.pdf

90 UN Economic Commission for Europe, 'UN Alliance aims to put fashion on path to sustainability', 12 July 2018. tinyurl.com/ja6n4yk8

91 European Parliament, 'The impact of textile production and waste on the environment'. tinyurl.com/jz5bsnjm

92 The *Guardian*, 'Child labourers exposed to toxic chemicals dying before 50, WHO says', 21 March 2017. tinyurl.com/25fkx9dh

93 UN Environment Programme, 'Global chemicals outlook: towards sound management of chemicals' (2013). tinyurl.com/53yayrmm

94 Greenpeace International, 'Dirty laundry: reloaded' (2012). tinyurl.com/f97z99z4

95 UN Environment Programme, 'Putting the brakes on fast fashion' (2018). tinyurl.com/b3d8epax

96 K. Niinimäki, G. Peters, H. Dahlbo *et al.*, 'The environmental price of fast fashion', *Nature Reviews Earth & Environment* (2020). doi.org/10.1038/s43017-020-0039-9

97 Fraunhofer Institute for Environmental, Safety and Energy Technology in Germany, 'Plastics in the environment: micro- and macroplastic' (2018). tinyurl.com/2efzb74p

98 The Danish Environmental Protection Agency, 'Polyfluoroalkyl substances (PFASs) in textiles for children' (2015). tinyurl.com/zds3fm75

99 Phil Patterson, 'Chemical circularity in fashion', commissioned by the Laudes Foundation (2020). tinyurl.com/3tjkbfxh
100 UK government consultation outcome (2019). tinyurl.com/z88phynv
101 National Ocean and Atmospheric Administration, US Department of Commerce, 'What is a dead zone?'. tinyurl.com/vcm64n4
102 Phoebe Racine, AnnaClaire Marley, Halley E. Froehlich *et al.*, 'A case for seaweed aquaculture inclusion in US nutrient pollution management', *Marine Policy* (2021). doi.org/10.1016/j.marpol.2021.104506
103 US Environmental Protection Agency podcast. tinyurl.com/4f2czdjd
104 Sierra Club, 'Sludge in the garden' (2021). tinyurl.com/yvty4fxs
105 Greenpeace, Unearthed and Public Eye investigation (2020). tinyurl.com/rech6ssw
106 Gregory S. Okin, 'Environmental impacts of food consumption by dogs and cats', *PLoS ONE* (2017). tinyurl.com/3757szjs
107 The Clean Label Project, 'Pet food study results 2017'. tinyurl.com/fjfks62j
108 Rajendiran Karthikraj, Sonali Borkar, Sunmi Lee and Kurunthachalam Kannan, 'Parabens and their metabolites in pet food and urine from New York State, United States', *Environmental Science & Technology* (2018). 10.1021/acs.est.7b05981
109 Liliana Finisterra, Bárbara Duarte, Luísa Peixe, Carla Novais and Ana R. Freitas, 'Industrial dog food is a vehicle of multidrug-resistant enterococci carrying virulence genes often linked to human infections', *International Journal of Food Microbiology* (2021). doi.org/10.1016/j.ijfoodmicro.2021.109284
110 Rosemary Perkins, Martin Whitehead, Wayne Civil and Dave Goulson, 'Potential role of veterinary flea products in widespread pesticide contamination of English rivers',

Science of The Total Environment (2021). doi.org/10.1016/j.scitotenv.2020.143560

111 J. H. Johansson *et al.*, 'Global transport of perfluoroalkyl acids via sea spray aerosol', *Environmental Science: Processes & Impacts* (2019). tinyurl.com/26tpu7sd

112 European Environment Agency, 'Chemicals'. eea.europa.eu/themes/human/chemicals

113 Kostyantyn Pivnenko, David Laner and Thomas F. Astrup, 'Material cycles and chemicals: dynamic material flow analysis of contaminants in paper recycling', *Environmental Science & Technology* (2016). doi.org/10.1021/acs.est.6b01791

114 Rosa María Espinosa Valdemar, Sylvie Turpin-Marion, Irma Delfín Alcalá and Alethia Vásquez Morillas, 'Disposable diapers biodegradation by the fungus *Pleurotus ostreatus*', *Waste Management* (2011). doi.org/10.1016/j.wasman.2011.03.007

Index

G

H

Q

R